Springer Texts in Business and Economics

More information about this series at
http://www.springer.com/series/10099

Igor V. Evstigneev • Thorsten Hens •
Klaus Reiner Schenk-Hoppé

Mathematical Financial Economics

A Basic Introduction

Igor V. Evstigneev
Economics, School of Social Sciences
University of Manchester
Manchester
United Kingdom

Thorsten Hens
Department of Banking and Finance
University of Zurich
Zurich
Switzerland

Klaus Reiner Schenk-Hoppé
Economics, School of Social Sciences
University of Manchester
Manchester
United Kingdom

ISSN 2192-4333 ISSN 2192-4341 (electronic)
Springer Texts in Business and Economics
ISBN 978-3-319-16570-7 ISBN 978-3-319-16571-4 (eBook)
DOI 10.1007/978-3-319-16571-4

Library of Congress Control Number: 2015939236

Springer Cham Heidelberg New York Dordrecht London

Printed on acid-free paper

Springer International Publishing AG Switzerland is part of Springer Science+Business Media (www.springer.com)

Preface

Tyche
Goddess of Chance and Fortune
By Tatjana Heinz

This textbook is a basic introduction to the key topics in mathematical finance and financial economics—two realms of ideas that substantially overlap but are often treated separately from each other. Our goal is to present the highlights in the field, with the emphasis on the financial and economic content of the models, concepts and results. The book provides a novel, unified treatment of the subject by deriving each topic from common fundamental principles and showing the interrelations between the key themes.

Although our presentation is fully rigorous, with some rare and clearly marked exceptions, we restrict ourselves to the use of only elementary mathematical concepts and techniques. No advanced mathematics (such as stochastic calculus) is used. The main source for the book, and a "proving ground" for testing our presentation of the material, are courses on mathematical finance, financial economics and risk management which we have delivered, over the last decade, to undergraduate and graduate students in economics and finance at the Universities of Manchester, Zurich and Leeds.

The textbook contains 18 chapters corresponding to 18 lectures in a course based upon it. There are three chapters with problems and exercises, most of which have been used in tutorials, take-home tests and examinations, with full and detailed answers. The problems and exercises contain not only numerical examples, but also theoretical questions that complement the material presented in the body of the textbook. Two mathematical appendices provide rigorous definitions of some of the mathematical notions and statements of general theorems used in the text.

The textbook covers the classical topics, such as mean-variance portfolio analysis (Markowitz, CAPM, factor models, the Ross-Huberman APT), derivative securities pricing, and general equilibrium models of asset markets (Arrow, Debreu and Radner). A less standard but very important topic, which to our knowledge has not previously been covered in elementary textbooks, is capital growth theory (Kelly, Breiman, Cover and others). Absolutely new material, reflecting research achievements of recent years, is an introduction to new dynamic equilibrium models of financial markets combining behavioral and evolutionary principles.

A characteristic feature of financial economics is that it has to focus on the analysis of random, unpredictable market situations. To model such situations our discipline created powerful theoretical tools based on probability and stochastic processes. However, the power of human mind is not unlimited, and it can never fully eliminate the influence of chance and fortune, personified by goddess Tyche, looking at us from the epigraph to this book.

Manchester, UK — Igor V. Evstigneev
Zurich, Switzerland — Thorsten Hens
Manchester, UK — Klaus Reiner Schenk-Hoppé

Contents

Part III Growth and Equilibrium

Part I

Mean-Variance Portfolio Analysis

1 Portfolio Selection: Introductory Comments

1.1 Asset Prices and Returns

Assets We will consider a financial market where N *assets* (*securities*) $i = 1, 2, \ldots, N$ are traded. Typical assets are common stocks, bonds, domestic or foreign cash, etc. Generally, the term "asset" is associated with any financial instrument that can be bought or sold.

Return Each asset is characterized by its *return* R_i. The return R_i on asset i is a random variable. For the purposes of mathematical modelling, we will assume that some characteristics of the random variables R_i, $i = 1, 2, \ldots, N$, (e.g. expectations and covariances) are known.

Asset Prices and Returns How are asset returns computed? We consider a model in which there are two moments of time 0 and 1. Let $S_0^i > 0$ be the price of asset i at time 0 and $S_1^i \geq 0$ the price of the asset at time 1. Then the *asset return* can be defined as

$$R_i = \frac{S_1^i - S_0^i}{S_0^i}.$$

This expression is also termed *the rate of return*.

Vectors of Prices and Vectors of Returns The financial market under consideration is specified by a random vector

$$R = (R_1, \ldots, R_N),$$

I.V. Evstigneev et al., *Mathematical Financial Economics*, Springer Texts in Business and Economics, DOI 10.1007/978-3-319-16571-4_1

whose ith component R_i represents the return on asset i. In what follows, we will denote by S_0 and S_1 the price vectors

$$S_0 = (S_0^1, \ldots, S_0^N), \; S_1 = (S_1^1, \ldots, S_1^N).$$

The price vector S_1 is random (not known at time 0, but known at time 1), while S_0 is fixed (known at time 0).

1.2 Investor's Portfolio: Long and Short Positions

Investor's Portfolio The problem of an investor is to decide what amount of what asset to buy, or in other words, what *portfolio* of assets to select. A portfolio x can be characterized by a vector

$$x = (x_1, \ldots, x_N),$$

where x_i denotes the amount of money invested in asset i. Assets are purchased at time 0, when their prices are $S_0^1, \ldots, S_0^N$. Total wealth invested in the portfolio $x = (x_1, \ldots, x_N)$ at time 0 is

$$w_0 = x_1 + \ldots + x_N.$$

For each i, the ratio

$$h_i = \frac{x_i}{S_0^i}$$

is the number of ("physical") units of asset i in the portfolio $x = (x_1, \ldots, x_N)$. At time 0, one can equivalently specify a portfolio in terms of the vector $(h_1, \ldots, h_N)$ or in terms of the vector $(x_1, \ldots, x_N)$.

Long and Short Positions of a Portfolio *Positions* of a portfolio (expressed in terms of money or in terms of physical units of assets) are the coordinates of the corresponding vector $x = (x_1, \ldots, x_N)$ or $h = (h_1, \ldots, h_N)$. These coordinates generally might be positive (*long* positions) or negative (*short* positions). A positive coordinate x_i = €100 of the vector x means that the investor owns the amount of asset i that costs (at time 0) €100. If, for example, S_0^i = €20, then the fact that x_i = €100 means that the investor owns $x_i / S_0^i = 100/20 = 5$ units of asset i.

What Does a Negative Coordinate x_i Mean? Negative positions of a portfolio appear, in particular, in the following three cases.

(a) If one of the assets, say $i = 1$, is cash, the negative number $x_1 = -€10{,}000$ means that the investor has *borrowed* €10,000 (for example, from a bank) at time 0. It is supposed that at time 1, the investor has to pay the debt $€(1 + r)10{,}000$, where $r \geq 0$ is the interest rate.
(b) Negative coordinates x_i of the portfolio might reflect a possibility of *short selling*. An investor may be allowed to borrow some amount of asset i from somebody (say, a broker) and sell this amount on the market at the prevailing price. At a later date, however, the assets must be returned. This may lead either to gains or to losses, depending on whether the price of the asset has decreased or increased during the time period under consideration. Short selling is a risky operation, and, in real financial institutions, it is often prohibited, or at least restricted by certain regulations.
(c) Suppose asset i is given by a contract, typically these are of some standardized form (an important example is an *option*; such contracts will be considered in detail later). Contracts of this kind may be written, sold and purchased by market traders. The one who writes contract i at time 0 can sell it at time 0 at a fixed price S_0^i, but must pay at time 1 some specified amount S_1^i (contingent on the random situation in the future) to the one who has purchased the contract. A trader who has written, say, 15 standardized contracts of type i and sold them at the current price S_0^i at time 0, must pay $15 \cdot S_1^i$ at time 1.

In this example, the ith position of the portfolio at time 0 is $-15 \cdot S_0^i$, and the ith position of the portfolio at time 1 is $-15 \cdot S_1^i$. The trader who has written the contract gains

$$(-15 \cdot S_1^i) - (-15 \cdot S_0^i) = 15 \cdot (S_0^i - S_1^i)$$

if $S_0^i - S_1^i > 0$ and loses this amount if $S_0^i - S_1^i < 0$. Such transactions might be motivated by the difference of the subjective expectations of the seller and the buyer of contract i. The former expects that $S_0^i - S_1^i$ will be positive (the price goes down), while the latter hopes it will be negative (the price goes up).

1.3 Return on a Portfolio

Initial and Terminal Values of a Portfolio Suppose that, at time 0, an investor constructs a portfolio $x = (x_1, x_2, \ldots, x_N)$, i.e., invests the amount x_i in asset i. The initial wealth w_0 of the investor, or the *initial value* of the portfolio (at time 0) is equal to the sum

$$w_0 = x_1 + x_2 + \ldots + x_N.$$

Since x_i is invested in asset i, the number of units of asset i in the investor's portfolio is

$$h_i = \frac{x_i}{S_0^i}.$$

The *terminal value* (at time 1) of this portfolio is

$$w_1 = \sum_{i=1}^{N} S_1^i h_i = \sum_{i=1}^{N} \frac{S_1^i x_i}{S_0^i} = \sum_{i=1}^{N} \frac{S_1^i - S_0^i}{S_0^i} x_i + \sum_{i=1}^{N} x_i = \left(\sum_{i=1}^{N} R_i x_i \right) + w_0.$$

Thus we obtained

$$w_1 = w_0 + \sum_{i=1}^{N} R_i x_i.$$

Consequently, the difference $w_1 - w_0$ between the terminal and initial values of the portfolio (capital gain) can be expressed as follows:

$$w_1 - w_0 = \sum_{i=1}^{N} R_i x_i. \tag{1.1}$$

Normalized Portfolios An investor is usually interested in the question: in what *proportions* should an amount of money $w_0 > 0$ be distributed between the assets $i = 1, 2, \ldots, N$? To analyze this question it is sufficient to consider portfolios $x = (x_1, \ldots, x_N)$ for which

$$x_1 + \ldots + x_N = 1.$$

Such portfolios are called *normalized.*

Negative Proportions? The term "proportions" mentioned above should be used with caution. Usually, this term is associated with numbers $p_1, \ldots, p_N$ such that $p_i \geq 0$ and $\sum p_i = 1$. In the present context, we do not assume that the proportions $x_1, \ldots, x_N$ of wealth invested in assets $i = 1, \ldots, N$ (positions of a normalized portfolio) are necessarily non-negative. Portfolio positions might be long or short, and so the numbers x_i might have positive and negative signs. However, the sum of these numbers, as long as they are regarded as proportions, is always equal to one.

Return on a Normalized Portfolio The main focus of the theory of portfolio selection is on investment proportions. Hence we will basically deal with normalized portfolios. For a normalized portfolio $x = (x_1, \ldots, x_N)$ its *return* is defined as

$$\frac{w_1 - w_0}{w_0} \tag{1.2}$$

or

$$\sum_{i=1}^{N} R_i x_i. \tag{1.3}$$

These numbers *coincide*, as long as the portfolio x is normalized. Indeed, then $w_0 = 1$, and so

$$\frac{w_1 - w_0}{w_0} = w_1 - w_0 = \sum_{i=1}^{N} R_i x_i$$

[see (1.1)].

Return on a Portfolio: The General Case For a general (not necessarily normalized) portfolio, the numbers (1.2) and (1.3) might be different. We will associate the term "*return*" on a portfolio $x = (x_1, \ldots, x_N)$ with the number defined by (1.3). To emphasize the distinction between (1.2) and (1.3) in the general case, we will call (1.2) the *net return* on the portfolio x. Note that (1.2) is defined only if $w_0 \neq 0$ (one cannot divide by $w_0 = 0$), while (1.3) is defined always.

Consider the simplest possible portfolio:

$$e_j = (0, 0, \ldots 0, 1, 0, \ldots 0).$$

(Here and in what follows, e_j stands for the vector whose coordinates are equal to 0, except for the jth coordinate which is equal to 1.) This portfolio does not contain any assets except j, the holding of this asset being worth €1. This portfolio is normalized, hence its net return is equal to its return, and we have:

$$x = e_j \Rightarrow \sum_{i=1}^{N} R_i x_i = R_j.$$

Thus the return on the portfolio e_j is equal to R_j, the return on asset j.

Computing Net Return For a portfolio $x = (x_1, \ldots, x_N)$ with $w_0 = \sum_i x_i \neq 0$, the net return equals

$$\frac{w_1 - w_0}{w_0} = \frac{\sum_i R_i x_i}{\sum_i x_i} = \sum_i R_i w_0^i,$$

where

$$w_0^i = \frac{x_i}{\sum_j x_j}$$

is the proportion of wealth (at time 0) invested in asset i.

Thus the *net return on a portfolio depends only on the proportions* of wealth invested in different assets.

Self-Financing Portfolios Portfolios $x = (x_1, \ldots, x_N)$ with zero initial value

$$x_1 + \ldots + x_N = 0$$

are called *self-financing*. How can these portfolios be created? For example, an investor can borrow some amount of money from a bank and buy some positive amounts of all the other assets. Then, if the first position x_1 of a portfolio $x = (x_1, \ldots, x_N)$ describes the investor's bank account, this position will be negative. The other positions will be positive, and the sum $\sum x_i$ will be zero. Clearly, in addition to borrowing, the investor may use the operation of short selling, which is permitted in the idealized market under consideration (but not always permitted in real markets). Note that the return

$$\sum_i x_i R_i$$

on a self-financing portfolio $(x_1, \ldots, x_N)$ is well-defined, while the net return $(w_1 - w_0)/w_0$ is not (because $w_0 = 0$). Since for a self-financing portfolio the initial value w_0 is equal to zero, we obtain

$$w_1 - w_0 = w_1, \tag{1.4}$$

and so *the return on a self-financing portfolio is equal to its terminal value* w_1.

1.4 Mathematical Notation

Notation: Scalar Product A mathematical comment is in order. According to our main definition, the return on a portfolio $x = (x_1, \ldots, x_N)$ is given by the formula

$$R_x = \sum_{i=1}^{N} x_i R_i .$$

This expression is nothing but the *scalar product* of two N-dimensional vectors

$$x = (x_1, \ldots, x_N) \text{ and } R = (R_1, \ldots, R_N).$$

(The former is deterministic, while the latter is random.)

In mathematics, several different symbols are used to denote the scalar product. In different contexts, it is convenient to use different symbols. We will use two different forms of notation. The scalar product of vectors $a = (a_1, \ldots, a_N)$ and $b = (b_1, \ldots, b_N)$ will be denoted either as ab or as $\langle a, b \rangle$. Both symbols have the same meaning

$$ab = \langle a, b \rangle = a_1 b_1 + \ldots + a_N b_N.$$

Sometimes the former notation will be more convenient, sometimes the latter.

2 Mean-Variance Portfolio Analysis: The Markowitz Model

2.1 Basic Notions

The Markowitz model[1] describes a market with N assets characterized by a random vector of returns

$$R = (R_1, \ldots, R_N).$$

The following data are assumed to be given:

- The expected value (mean) $m_i = ER_i$ of each random variable R_i, $i = 1, 2, \ldots, N$;
- The covariances $\sigma_{ij} = Cov(R_i, R_j)$ for all pairs of random variables R_i and R_j.

The covariance of two random variables, X and Y, is defined by

$$Cov(X, Y) = E[X - EX][Y - EY] = E(XY) - (EX)(EY).$$

We will denote by m the vector of the expected returns

$$m = (m_1, \ldots, m_N)$$

and by V the covariance matrix

$$V = (\sigma_{ij}), \ \sigma_{ij} = Cov(R_i, R_j)$$

[1] Markowitz, H., Portfolio Selection, Journal of Finance 7, 77–91, 1952. Markowitz was awarded a Nobel Prize in Economics in 1990, jointly with W. Sharpe and M. Miller.

I.V. Evstigneev et al., *Mathematical Financial Economics*, Springer Texts in Business and Economics, DOI 10.1007/978-3-319-16571-4_2

of the random vector $R = (R_1, \ldots, R_N)$. (The expectations and the covariances are assumed to be well-defined and finite.) The matrix V has N rows and N columns. The element at the intersection of ith row and jth column is σ_{ij}.

Expectations and Covariances of Returns Consider a portfolio $x = (x_1, \ldots, x_N)$, where x_i is the amount of money invested in asset i. Recall that the return on the portfolio x is computed according to the formula

$$R_x = \sum_{i=1}^{N} x_i R_i.$$

Consequently, the expected return $m_x = ER_x$ on the portfolio x is given by

$$m_x = \sum_{i=1}^{N} x_i m_i = \langle m, x \rangle$$

where

$$m_i = ER_i$$

and

$$m = (m_1, \ldots, m_N).$$

The variance $VarR_x$ of the portfolio return R_x can be computed as follows:

$$\begin{aligned}
\sigma_x^2 &= Var(R_x) = E(R_x - m_x)^2 \\
&= E\left(\sum_{i=1}^{N} x_i R_i - \sum_{i=1}^{N} x_i m_i\right)^2 = E\left[\sum_{i=1}^{N} x_i (R_i - m_i)\right]^2 \\
&= E\left[\sum_{i=1}^{N} x_i (R_i - m_i)\right]\left[\sum_{j=1}^{N} x_j (R_j - m_j)\right] \\
&= E\left[\sum_{i=1}^{N}\sum_{j=1}^{N} x_i x_j (R_i - ER_i)(R_j - ER_j)\right] \\
&= \sum_{i=1}^{N}\sum_{j=1}^{N} x_i Cov(R_i, R_j) x_j \\
&= \sum_{i=1}^{N}\sum_{j=1}^{N} x_i \sigma_{ij} x_j = \langle x, Vx \rangle.
\end{aligned}$$

Thus we have the following formulas for the expectation and the variance of the return R_x on the portfolio x:

$$m_x = ER_x = \langle m, x \rangle, \tag{2.1}$$

$$\sigma_x^2 = Var(R_x) = \langle x, Vx \rangle. \tag{2.2}$$

Markowitz's Approach to Portfolio Selection This approach is often used in practical decisions. Given the constraint $\sum x_i = 1$ on the portfolio weights, investors choose a portfolio x, having two objectives:

- Maximization of the expected value $m_x = ER_x$ of the portfolio return;
- Minimization of the portfolio *risk*, which is measured by $\sigma_x^2 = VarR_x$ or σ_x.

We denote by σ_x the *standard deviation* of the random variable R_x:

$$\sigma_x = \sqrt{VarR_x} = \sqrt{E(R_x - m_x)^2}.$$

It is the fundamental assumption of the Markowitz approach that only two numbers characterize the portfolio: the expectation and the variance of the portfolio return. The variance is used as a very simple measure of risk: the more "variable" the random return R_x on the portfolio x, the higher the variance of R_x. If the return R_x is certain, its variance is equal to zero, and so such a portfolio is *risk-free*.

2.2 Optimization Problem: Formulation and Discussion

The Markowitz Optimization Problem According to individual preferences, an investor puts weights on the conflicting objectives m_x and σ_x^2 and maximizes

$$\tau m_x - \sigma_x^2$$

given the parameter $\tau \geq 0$. This parameter is called *risk tolerance*. Hence, according to Markowitz, the optimization problem to be solved is as follows:

$$\max_{x \in R^N} \{\tau m_x - \sigma_x^2\}$$

subject to

$$x_1 + \ldots + x_N = 1.$$

More explicitly, the above problem can be written

$$\max_{x \in R^N} \left\{ \tau \sum_{i=1}^{N} m_i x_i - \sum_{i=1}^{N} \sum_{j=1}^{N} x_i \sigma_{ij} x_j \right\}$$

subject to

$$x_1 + \ldots + x_N = 1.$$

Using the notation

$$e = (1, 1, \ldots, 1)$$

for the vector whose all coordinates are equal to one and writing $\langle \cdot, \cdot \rangle$ for the scalar product, we can represent the Markowitz optimization problem as follows:

$$\max_{x \in R^N} \{ \tau \langle m, x \rangle - \langle x, Vx \rangle \}$$

subject to

$$\langle e, x \rangle = 1.$$

Advantages and Disadvantages of the Markowitz Approach The Markowitz approach has the following important *advantages*:

- The preferences of the investor are described in a most simple way. Only one positive number, the risk tolerance τ, has to be determined.
- Only the expectations $m_i = ER_i$ and the covariances $\sigma_{ij} = Cov(R_i, R_j)$ of asset returns are needed.
- The optimization problem is quadratic concave, and powerful numerical algorithms exist for finding its solutions.
- Most importantly, the Markowitz optimization problem admits an explicit analytic solution, which makes it possible to examine its quantitative and qualitative properties in much detail.

The main *drawback* of the Markowitz approach is its inability to cover situations in which the distribution of the portfolio return cannot be fully characterized by such a scarce set of data as m_i and σ_{ij}.

Efficient Portfolios Portfolios obtained by using the Markowitz approach are termed *efficient*.

Definition A portfolio x^* is called *(mean-variance) efficient* if it solves the optimization problem

$$(\mathbf{M}_\tau) \qquad \max_{x \in R^N} \{\tau m_x - \sigma_x^2\}$$

$$\text{subject to: } x_1 + \ldots + x_N = 1$$

for some $\tau \geq 0$.

2.3 Assumptions

Basic Assumptions We will start the analysis of the Markowitz model under the following assumptions (later, an alternative set of assumptions will be considered).

Assumption 1 The covariance matrix V is *positive definite*.

This assumption means that

$$\langle x, Vx \rangle \left(= \sum_{i,j=1}^{N} x_i \sigma_{ij} x_j \right) > 0 \text{ for each } x \neq 0.$$

Since $\langle x, Vx \rangle = Var(R_x)$, we always have $\langle x, Vx \rangle \geq 0$. The above assumption requires that $\langle x, Vx \rangle = 0$ *only if* $x = 0$. As a consequence of Assumption 1, we obtain $VarR_i > 0$, i.e., *all the assets* $i = 1, 2, \ldots, N$ *are risky*.

If Assumption 1 is satisfied, then the objective function

$$\tau m_x - \sigma_x^2 = \tau \langle m, x \rangle - \langle x, Vx \rangle$$

in the Markowitz problem $(\mathbf{M}_\tau)$ is strictly concave and the solution to $(\mathbf{M}_\tau)$ exists and is unique.[2]

The set of efficient portfolios is a one-parameter family with parameter τ ranging through the set $[0, \infty)$ of all non-negative numbers.

The efficient portfolio x^{MIN} corresponding to $\tau = 0$ is termed the *minimum variance portfolio*. It minimizes $VarR_x = \langle x, Vx \rangle$ over all normalized portfolios x.

What Happens If Assumption 1 Fails to Hold? Then there is a portfolio $y \neq 0$ with $\langle y, Vy \rangle = 0$. Hence

$$Var(R_y) = Var(y_1 R_1 + \ldots + y_N R_N) = 0.$$

[2]For details see Mathematical Appendix A.

Thus R_y is equal to a constant, c, with probability one. If $c \neq 0$, we can assume without loss of generality that $c > 0$ (replace y by $-y$ if needed!). The property

$$y_1 R_1 + \ldots + y_N R_N = c > 0 \text{ with probability } 1$$

means the existence of a *risk-free investment strategy with strictly positive return* (which is ruled out in the present context).

If $c = 0$, then the equality $y_1 R_1 + \ldots + y_N R_N = 0$, holding for some $(y_1, \ldots, y_N) \neq 0$, means that the random variables $R_1, \ldots, R_N$ are *linearly dependent*. Then at least one of them (any one for which $y_i \neq 0$) can be expressed as a linear combination of the others, which means the existence of a *redundant asset*.

In addition to Assumption 1, we will need the following

Assumption 2 There are at least two assets i and j with expected returns $m_i \neq m_j$.

What If Assumption 2 Does Not Hold? If Assumption 2 is not satisfied, then there is only one efficient portfolio, x^{MIN}. Indeed, if Assumption 2 does not hold, then all the numbers $m_1, \ldots, m_N$ are the same and are equal, say, to some number θ. Then we have $m = \theta e$, i.e., the vectors m and $e = (1, 1, \ldots, 1)$ are collinear. In the Markowitz problem $(\mathbf{M}_\tau)$, we have to maximize

$$\tau \langle m, x \rangle - \langle x, Vx \rangle$$

under the constraint

$$\langle e, x \rangle = 1.$$

If $m = \theta e$, then for every x satisfying the constraint $\langle e, x \rangle = 1$, the value of the objective function is equal to

$$\tau \langle m, x \rangle - \langle x, Vx \rangle = \tau\theta \langle e, x \rangle - \langle x, Vx \rangle = \tau\theta - \langle x, Vx \rangle.$$

For each τ, the maximum value of this function is attained at $x = x^{MIN}$ because x^{MIN} minimizes $\langle x, Vx \rangle$ on the set of all normalized portfolios.

2.4 Efficient Portfolios and Efficient Frontier

Efficient Frontier We can draw a diagram depicting the set of all points (σ_x^2, m_x) in the plane corresponding to all efficient portfolios x. This set is called the *efficient frontier*. The efficient frontier is a curve of the following typical form (Fig. 2.1):

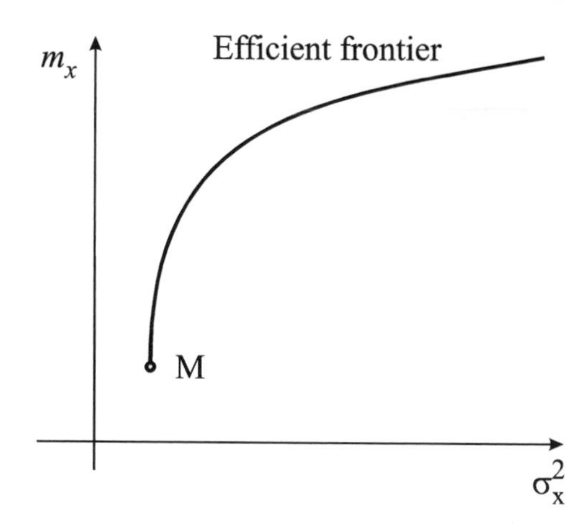

Fig. 2.1 Efficient frontier

The point M of the curve in the above diagram corresponds to the minimum variance efficient portfolio (for which $\tau = 0$). All the other points (σ_x^2, m_x) of the curve represent the variances and the expectations of the returns on efficient portfolios x with $\tau > 0$.

Efficient Portfolios: An Equivalent Definition We give an equivalent definition of an efficient portfolio (which is often used in the literature).

Proposition 2.1 *A normalized portfolio $x^* \in R^N$ is efficient if and only if there exists no normalized portfolio $x \in R^N$ such that*

$$m_x \geq m_{x^*} \text{ and } \sigma_x^2 < \sigma_{x^*}^2.$$

The last two inequalities mean that x^* solves the optimization problem

$$(\mathbf{M}^\mu) \qquad \min_{x \in R^N} \sigma_x^2$$

subject to

$$m_x \geq \mu \text{ and } \sum x_i = 1,$$

where $\mu = m_{x^*}$ and $x = (x_1, \ldots, x_N)$.

Proof "Only if": We have to show that if x^* is a solution to $(\mathbf{M}_\tau)$, then x^* is a solution to $(\mathbf{M}^\mu)$ with $\mu = m_{x^*}$. Suppose the contrary: x^* is a solution to $(\mathbf{M}_\tau)$, but not to $(\mathbf{M}^\mu)$, i.e., there is a normalized portfolio x for which $m_x \geq \mu = m_{x^*}$ and $\sigma_x^2 < \sigma_{x^*}^2$. Then $\tau m_x - \sigma_x^2 > \tau m_{x^*} - \sigma_{x^*}^2$, which means that x^* is *not* a solution to $(\mathbf{M}_\tau)$. A contradiction.

"If": We have to show that if x^* is a solution to $(\mathbf{M}^\mu)$ with $\mu = m_{x^*}$, then x^* is a solution to $(\mathbf{M}_\tau)$ for some $\tau \geq 0$. It can be shown that there exists a Lagrange multiplier $\gamma \geq 0$ relaxing the constraint $m_x \geq \mu$ in $(\mathbf{M}^\mu)$:

$$-\sigma_x^2 + \gamma(m_x - \mu) \leq -\sigma_{x^*}^2 + \gamma(m_{x^*} - \mu)$$

for each normalized portfolio x. This implies

$$\gamma m_x - \sigma_x^2 \leq \gamma m_{x^*} - \sigma_{x^*}^2.$$

By setting $\tau = \gamma$, we obtain that x^* is a solution to $(\mathbf{M}_\tau)$, which completes the proof. □

Remark The above proof is based on a general result on the existence of Lagrange multipliers for convex optimization problems—the Kuhn–Tucker theorem. This theorem is presented in Mathematical Appendix B.

Solution to the Markowitz Optimization Problem

3

3.1 Statement of the Main Result

Solving the Markowitz Problem As we have said, a remarkable feature of the Markowitz portfolio selection problem is that it admits an explicit analytic solution. In this chapter, we first formulate this solution, then discuss it and then prove the result.

Recall that the model deals with a financial market where N assets $i = 1, 2, \ldots, N$ are traded. Each asset i is characterized by a random variable R_i, this asset's return. Thus the market is characterized by a random vector of returns

$$R = (R_1, \ldots, R_N).$$

As in Chap. 2, the following data are given:

- The expected value (mean) $m_i = ER_i$ of each random variable R_i, $i = 1, 2, \ldots, N$;
- The covariances $\sigma_{ij} = Cov(R_i, R_j)$ for all pairs of random variables R_i and R_j.

Notation and Assumptions Let us recall our basic notation and assumptions. We denote by m the vector of the expected returns and by V the covariance matrix of the random vector $R = (R_1, \ldots, R_N)$:

$$m = (m_1, \ldots, m_N), \quad V = (\sigma_{ij}), \ \sigma_{ij} = Cov(R_i, R_j).$$

We impose the following assumptions (discussed in detail in the previous chapter):

Assumption 1 The covariance matrix V is *positive definite*.

I.V. Evstigneev et al., *Mathematical Financial Economics*, Springer Texts in Business and Economics, DOI 10.1007/978-3-319-16571-4_3

Assumption 2 There are at least two assets i and j with $m_i \neq m_j$.

Recall that for each portfolio $x = (x_1, \ldots, x_N)$, the expectation $m_x = ER_x$ and the variance $\sigma_x^2 = VarR_x$ of the portfolio return

$$R_x = x_1 R_1 + \ldots + x_N R_N = \langle x, R \rangle$$

are given by

$$m_x = \langle x, m \rangle \text{ and } \sigma_x^2 = \langle x, Vx \rangle.$$

Formulation of the Markowitz Optimization Problem An investor with *risk tolerance* $\tau \geq 0$ solves the following optimization problem:

$$(\mathbf{M}_\tau) \qquad \max_{x \in R^N} \{\tau m_x - \sigma_x^2\}$$

subject to

$$x_1 + \ldots + x_N = 1.$$

We always denote by e the vector $e = (1, 1, \ldots, 1) \in R^N$. Then the Markowitz problem can be written

$$(\mathbf{M}_\tau) \qquad \max_{x \in R^N} \{\tau \langle m, x \rangle - \langle x, Vx \rangle\} \tag{3.1}$$

subject to

$$\langle x, e \rangle = 1. \tag{3.2}$$

Statement of the Result We provide an *explicit solution* to problem $(\mathbf{M}_\tau)$ under Assumptions 1 and 2. The solution is given in terms of the inverse[1] of the covariance matrix V, which will be denoted by W:

$$W = V^{-1}.$$

Theorem 3.1 *For each $\tau \geq 0$ the Markowitz portfolio selection problem $(\mathbf{M}_\tau)$ has a unique solution x_τ^* given by*

$$x_\tau^* = x^{MIN} + \frac{\tau}{2} z^*, \tag{3.3}$$

[1] By the definition of an inverse matrix, we have $WV = VW = Id$, where Id is the identity matrix. The inverse matrix V^{-1} exists when the equation $Vx = 0$ has the unique solution $x = 0$ (e.g. if V is positive definite).

where

$$x^{MIN} = \frac{We}{\langle e, We\rangle} \; (= x_0^*) \tag{3.4}$$

and

$$z^* = Wm - \frac{\langle e, Wm\rangle}{\langle e, We\rangle} We. \tag{3.5}$$

3.2 Discussion

Several Remarks Before proving the theorem, let us discuss the result. According to the theorem, the optimal portfolio can be represented in the form (3.3).

Remark 3.1 First of all, observe that the vector x^{MIN} is the solution to the Markowitz problem $(\mathbf{M}_\tau)$ corresponding to $\tau = 0$. When $\tau = 0$, the objective function is equal to $-\sigma_x^2$, and so the portfolio x^{MIN} solves the following problem:

$$(\mathbf{M}_0) \qquad \min_{x \in R^N} \sigma_x^2$$

subject to

$$\langle x, e\rangle = 1.$$

Thus x^{MIN} minimizes the variance of return among all normalized portfolios x. In this connection, x^{MIN} is called the *minimum variance portfolio.*

The fact that the portfolio given by (3.4) is normalized, i.e. satisfies $\langle e, x^{MIN}\rangle = 1$ (as it really should be) can be deduced directly from formula (3.4):

$$\langle e, x^{MIN}\rangle = \langle e, \frac{We}{\langle e, We\rangle}\rangle = \frac{\langle e, We\rangle}{\langle e, We\rangle} = 1.$$

Remark 3.2 Observe that the portfolio z^* defined by (3.5) satisfies

$$\langle e, z^*\rangle = \langle e, Wm\rangle - \frac{\langle e, Wm\rangle}{\langle e, We\rangle}\langle e, We\rangle = 0,$$

and so $\langle e, z^*\rangle = \sum z_i^* = 0$. Hence z^* is a *self-financing* portfolio.

Remark 3.3 Let us compute the expected return $m^*_\tau = m_{x^*_\tau}$ of the optimal portfolio x^*_τ. We have

$$m_{x^*_\tau} = \langle m, x^{MIN} \rangle + \frac{\tau}{2} \langle m, z^* \rangle,$$

where

$$\langle m, z^* \rangle = \langle m, Wm \rangle - \frac{\langle e, Wm \rangle}{\langle e, We \rangle} \langle m, We \rangle$$

$$= \frac{\langle m, Wm \rangle \langle e, We \rangle - \langle e, Wm \rangle^2}{\langle e, We \rangle}. \tag{3.6}$$

To obtain the last formula, we used the following fact:

$$\langle e, Wm \rangle = \langle m, We \rangle.$$

This is so because the matrix $W = V^{-1}$ is *symmetric* ($w_{ij} = w_{ji}$), which, in turn, follows from the fact that *the covariance matrix* $V = (\sigma_{ij})$ *is symmetric*: $\sigma_{ij} = Cov(R_i, R_j) = Cov(R_j, R_i) = \sigma_{ji}$. (For details see Mathematical Appendix A.)

Remark 3.4 We obtained that the expected return of the portfolio z^* is given by (3.6). Let us show that this expression is strictly positive. Clearly $\langle e, We \rangle > 0$ because W is positive definite. It remains to verify that

$$\langle m, Wm \rangle \langle e, We \rangle > \langle e, Wm \rangle^2.$$

This follows from the *Cauchy–Schwartz inequality* (see Mathematical Appendix A). This inequality holds as long as the vectors e and m are not *collinear*, i.e., m cannot be represented as $m = \lambda e$ for some number λ. This is so by virtue of Assumption 2: $m_i \neq m_j$ for some i and j.

Remark 3.5 Thus we have shown that the expected return of the portfolio x^*_τ satisfies

$$ER_{x^*_\tau} = ER_{x^{MIN}} + \frac{\tau}{2} ER_{z^*}, \text{ where } ER_{z^*} > 0.$$

Therefore the formula

$$x^*_\tau = x^{MIN} + \frac{\tau}{2} z^* \tag{3.7}$$

for the portfolio solving the Markowitz problem ($\mathbf{M}_\tau$) may be interpreted as follows. The portfolio x^{MIN} leads to a minimum risk investment strategy. This strategy is corrected by the self-financing return-generating portfolio z^*. The investment

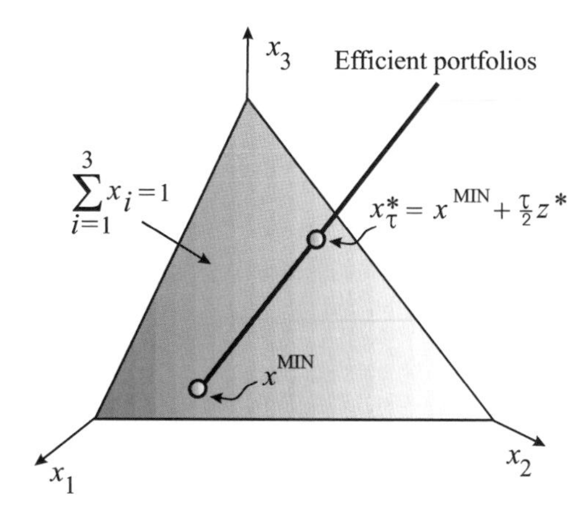

Fig. 3.1 The set of efficient portfolios

in each position of the portfolio z^* is proportional to the coefficient τ, the *risk tolerance*.

Geometrically, the set of efficient portfolios

$$x^*_\tau = x^{MIN} + \frac{\tau}{2} z^*, \ \tau \geq 0, \tag{3.8}$$

can be represented as a half-line (ray) emanating from the point x^{MIN} in the N-dimensional space R^N. Since all such portfolios are normalized, this line is contained in the hyperplane $\sum x_i = 1$. This situation is illustrated in Fig. 3.1 for $N = 3$.

3.3 Proof of the Main Result

Proof of Theorem 3.1

1st step. Consider the Markowitz problem ($\mathbf{M}_\tau$) described in (3.1) and (3.2). To solve ($\mathbf{M}_\tau$) we use the method of Lagrange multipliers. Consider the Lagrangian

$$L(x, \lambda) = \tau \langle m, x \rangle - \langle x, Vx \rangle + \lambda(\langle x, e \rangle - 1),$$

for the constraint $\langle x, e \rangle - 1 = 0$. We will find a vector x and a Lagrange multiplier λ for which the gradient L'_x (the vector of the partial derivatives of L with respect to x) is equal to zero:

$$\tau m - 2Vx + \lambda e = 0 \tag{3.9}$$

and the constraint

$$\langle x, e \rangle = 1 \tag{3.10}$$

holds. This vector $x = x_\tau^*$ will be the solution to $(\mathbf{M}_\tau)$.
We use here the following. The gradient of the function $\langle x, Vx \rangle$ is $2Vx$:

$$\left(\frac{\partial}{\partial x_1} \langle x, Vx \rangle, \ldots, \frac{\partial}{\partial x_N} \langle x, Vx \rangle \right) = 2Vx.$$

The Hessian of the function $\langle x, Vx \rangle$ is $2V$:

$$\left(\frac{\partial^2}{\partial x_i \partial x_j} \langle x, Vx \rangle \right) = 2V.$$

The Hessian is positive definite, and so the function $\langle x, Vx \rangle$ is strictly convex. (See Mathematical Appendices A and B.)

2nd step. To find $x = x_\tau^*$ we have to solve the system of two equations given by (3.9) and (3.10). Let us first solve the system for $\tau = 0$. This will give us the minimum variance portfolio $x^{MIN} = x_0^*$. By setting $\tau = 0$ and by applying the matrix $W = V^{-1}$ to Eq. (3.9), we get

$$-2x + \lambda We = 0,$$

which implies $-2\langle e, x \rangle + \lambda \langle e, We \rangle = 0$. Since $\langle e, x \rangle = 1$ [see (3.10)], we obtain

$$\lambda = \frac{2}{\langle e, We \rangle}.$$

Thus the solution x to (3.9) and (3.10) for $\tau = 0$ is as follows:

$$x^{MIN} = x_0^* = \frac{We}{\langle e, We \rangle}.$$

3rd step. To solve the system of equations (3.9) and (3.10) for any $\tau > 0$, we denote by z the difference

$$z = x - x^{MIN} = x - \frac{We}{\langle e, We \rangle}.$$

By substituting $x = z + \dfrac{We}{\langle e, We\rangle}$ into (3.9) and (3.10), we arrive at the system

$$\tau m - 2Vz - \frac{2e}{\langle e, We\rangle} + \lambda e = 0; \tag{3.11}$$

$$\langle z, e\rangle = 0. \tag{3.12}$$

By applying the matrix $W = V^{-1}$ to the first Eq. (3.11), we find

$$\tau Wm - 2z - \frac{2We}{\langle e, We\rangle} + \lambda We = 0. \tag{3.13}$$

The scalar product of this expression and e gives

$$\tau\langle e, Wm\rangle - 2 + \lambda\langle e, We\rangle = 0,$$

where we have used the fact that $\langle e, z\rangle = 0$ by virtue of (3.12). From the last formula, we obtain

$$\lambda = \frac{2}{\langle e, We\rangle} - \frac{\tau\,\langle e, Wm\rangle}{\langle e, We\rangle},$$

and so (3.13) becomes

$$\tau Wm - 2z - \frac{\tau\,\langle e, Wm\rangle}{\langle e, We\rangle} We = 0.$$

The last equation gives

$$z = \frac{\tau}{2}z^*, \text{ where } z^* = Wm - \frac{\langle e, Wm\rangle}{\langle e, We\rangle} We,$$

which completes the proof of Theorem 3.1.

□

4 Properties of Efficient Portfolios

4.1 Mean and Variance of the Return on an Efficient Portfolio

Analyzing the Solution to the Markowitz Problem We have shown in the previous chapter that the efficient portfolio x_τ^* solving the Markowitz problem $(\mathbf{M}_\tau)$ is given by

$$x_\tau^* = x^{MIN} + \frac{\tau}{2} z^*,$$

where

$$x^{MIN} = \frac{We}{\langle e, We \rangle}, \; z^* = Wm - \frac{\langle e, Wm \rangle}{\langle e, We \rangle} We.$$

Notation (The "Magic" Numbers A, B, C and D) Let us introduce the following convenient notation:

$$A = \langle e, Wm \rangle, \; B = \langle m, Wm \rangle, \; C = \langle e, We \rangle, \; D = BC - A^2.$$

Here $B > 0$, $C > 0$ because the matrix W is positive definite and $D > 0$ by virtue of the Cauchy–Schwartz inequality (see Appendix A). By using this notation, we can write

$$x^{MIN} = \frac{We}{C}, \; z^* = Wm - \frac{A}{C} We.$$

I.V. Evstigneev et al., *Mathematical Financial Economics*, Springer Texts in Business and Economics, DOI 10.1007/978-3-319-16571-4_4

The Expected Return on x_τ^* We have shown (see Remark 3.3) that the expected return on the portfolio x_τ^* is equal to

$$m_{x_\tau^*} = \langle m, x^{MIN} \rangle + \frac{\tau}{2} \langle m, z^* \rangle.$$

We now express $m_{x_\tau^*}$ through A, B, C and D:

$$\langle m, z^* \rangle = \langle m, Wm \rangle - \frac{A}{C} \langle m, We \rangle = \frac{BC - A^2}{C} = \frac{D}{C}$$

and

$$\langle m, x^{MIN} \rangle = \frac{\langle m, We \rangle}{C} = \frac{\langle Wm, e \rangle}{C} = \frac{A}{C}.$$

Thus we have obtained a formula for the **expected return** on x_τ^*

$$ER_{x_\tau^*} = \frac{A}{C} + \frac{\tau}{2} \frac{D}{C}.$$

The Variance of the Return on x_τ^* We have

$$R_{x_\tau^*} = R_{x^{MIN}} + \frac{\tau}{2} R_{z^*}.$$

To compute $VarR_{x_\tau^*}$ we proceed as follows.

1st step. We compute the variance of $R_{x^{MIN}}$:

$$VarR_{x^{MIN}} = \langle x^{MIN}, Vx^{MIN} \rangle = \langle \frac{We}{C}, V \frac{We}{C} \rangle = \langle \frac{We}{C}, \frac{e}{C} \rangle = \frac{1}{C}.$$

2nd step. By using formula (2.2) derived in Chap. 2, we compute the variance of R_{z^*}:

$$\begin{aligned}
VarR_{z^*} &= \langle Wm - \frac{A}{C} We, V(Wm - \frac{A}{C} We) \rangle \\
&= \langle Wm - \frac{A}{C} We, m - \frac{A}{C} e \rangle \\
&= \langle Wm, m \rangle - \frac{A}{C} \langle We, m \rangle - \frac{A}{C} \langle Wm, e \rangle + \frac{A^2}{C^2} \langle We, e \rangle \\
&= B - 2\frac{A^2}{C} + \frac{A^2}{C} = B - \frac{A^2}{C} = \frac{D}{C}.
\end{aligned}$$

3rd step. We observe that the random variables R_{z^*} and $R_{x^{MIN}}$ are *uncorrelated:*

$$Cov\langle R_{z^*}, R_{x^{MIN}}\rangle = \langle z^*, Vx^{MIN}\rangle = \frac{\langle z^*, e\rangle}{C} = 0$$

because, as we proved above, z^* is self-financing. We use here the formula

$$Cov\langle R_x, R_y\rangle = \langle x, Vy\rangle. \tag{4.1}$$

In Chap. 2, this formula was proved for $x = y$. Then $Cov\langle R_x, R_x\rangle = VarR_x = \langle x, Vx\rangle$. The derivation of (4.1) in the general case is similar to that in Chap. 2. We leave this derivation as an exercise to the reader (Hint: replace in all the formulas x_j by y_j).

4th step. If two random variables are non-correlated, then the sum of their variances is equal to the variance of their sum (verify!). Consequently, the **variance of the return** on x^*_τ can be expressed as

$$VarR_{x^*_\tau} = VarR_{x^{MIN}} + \frac{\tau^2}{4}VarR_{z^*} = \frac{1}{C} + \frac{\tau^2}{4}\frac{D}{C}.$$

Thus we have obtained the following **equations for the expectation and the variance of the return on the optimal portfolio** x^*_τ:

$$ER_{x^*_\tau} = \frac{A}{C} + \frac{\tau}{2}\frac{D}{C}, \quad VarR_{x^*_\tau} = \frac{1}{C} + \frac{\tau^2}{4}\frac{D}{C}. \tag{4.2}$$

4.2 Description of the Efficient Frontier

Efficient Frontier The result obtained gives a complete description of the *efficient frontier*. The efficient frontier (see Chap. 2) can be defined as the set of those points in the plane which correspond to the variances and the expectations of returns on efficient portfolios—solutions to the Markowitz problem ($\mathbf{M}_\tau$), where τ ranges through all non-negative numbers. Thus in the plane with coordinates σ_x^2, m_x, the efficient frontier is the curve consisting of the points (σ_x^2, m_x) where

$$\sigma_x^2 = \frac{1}{C} + \frac{\tau^2}{4}\frac{D}{C}, \quad m_x = \frac{A}{C} + \frac{\tau}{2}\frac{D}{C}, \quad \tau \geq 0.$$

Excluding τ from these equations, we obtain that the curve under consideration is (the upper portion of) the *parabola*

$$\sigma_x^2 = \frac{1}{D}(Cm_x^2 - 2Am_x + B)$$

shown in the diagram below (Fig. 4.1).

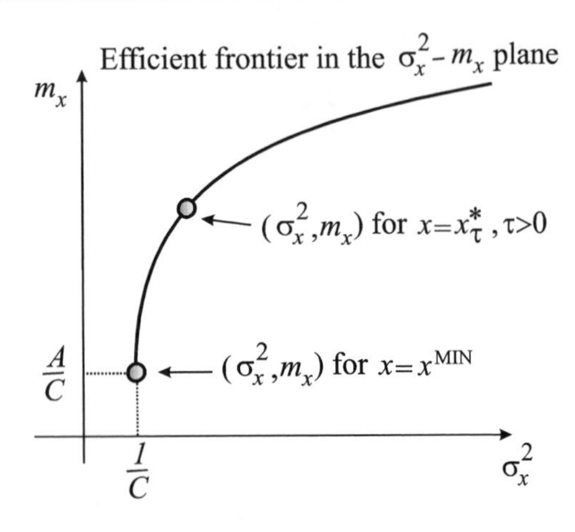

Fig. 4.1 Efficient frontier in the σ_x^2-m_x plane

The above figure depicts the efficient frontier in the σ_x^2-m_x plane. The coordinates of the points of the frontier curve are the variances and the expectations of the returns on efficient portfolios. We can also draw an analogous curve in the σ_x-m_x plane, where the coordinates are the standard deviations and the expectations of returns. *The efficient frontier in the σ_x-m_x plane* is (the upper portion of) the *hyperbola*

$$\sigma_x = \sqrt{\frac{1}{D}(Cm_x^2 - 2Am_x + B)}$$

with vertex at $(\sqrt{1/C}, A/C)$ and asymptote[1]

$$\sigma_x = A/C + \sqrt{C/D}\, m_x;$$

see Fig. 4.2.

4.3 A Fund Separation Theorem

The Two-Fund Theorem By virtue of Theorem 3.1, every efficient portfolio can be represented as

$$x_\tau^* = x^{MIN} + \frac{\tau}{2} z^*, \ \tau \geq 0.$$

Fix any two numbers $0 \leq \tau_1 < \tau_2$ and consider the efficient portfolios $x_{\tau_1}^*$ and $x_{\tau_2}^*$.

[1] An asymptote of a curve is a straight line that is tangent to the curve at infinity.

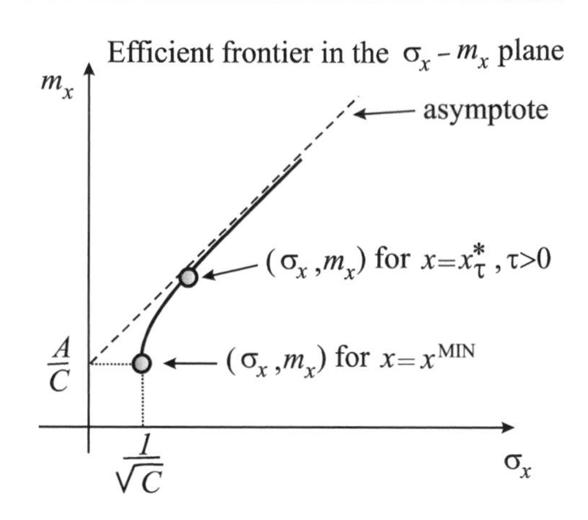

Fig. 4.2 Efficient frontier in the σ_x-m_x plane

Theorem 4.1 *Every efficient portfolio x can be represented as a combination*

$$x = \alpha x^*_{\tau_1} + (1-\alpha)x^*_{\tau_2}$$

of the portfolios $x^*_{\tau_1}$ and $x^*_{\tau_2}$ with some α.

Proof We know that every efficient portfolio x is of the form $x = x^*_\tau = x^{MIN} + \frac{\tau}{2}z^*$, where τ is some non-negative number. Define

$$\alpha = \frac{\tau - \tau_2}{\tau_1 - \tau_2}.$$

Then, as is easily seen,

$$\alpha\tau_1 + (1-\alpha)\tau_2 = \tau,$$

and so

$$\begin{aligned} &\alpha x^*_{\tau_1} + (1-\alpha)x^*_{\tau_2} \\ &= \alpha x^{MIN} + (1-\alpha)x^{MIN} + [\alpha\frac{\tau_1}{2} + (1-\alpha)\frac{\tau_2}{2}]z^* \\ &= x^{MIN} + \frac{\tau}{2}z^* = x^*_\tau = x, \end{aligned}$$

which completes the proof. □

The above result, which is a simple consequence of Theorem 3.1, is known as a *two-fund theorem*. It states that two efficient "mutual funds" (portfolios) can be established, so that any efficient portfolio can be represented as a combination of these two.

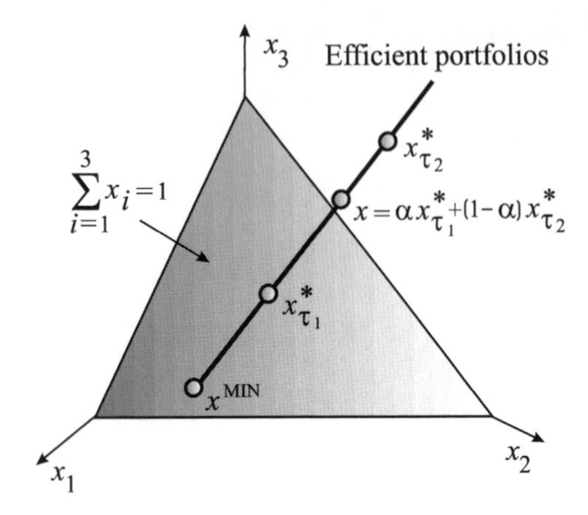

Fig. 4.3 The two-fund theorem: a geometric illustration

A Geometric Illustration It has been noticed that all the vectors representing efficient portfolios form a ray in the N-dimensional space. Geometrically, the assertion of Theorem 4.1 is clear: if we fix any two points (efficient portfolios) $x^*_{\tau_1}$ and $x^*_{\tau_2}$ on this ray, then any other point (efficient portfolio) x on this ray can be obtained as a weighted average

$$x = \alpha x^*_{\tau_1} + (1 - \alpha) x^*_{\tau_2}$$

of these two points. The weights α and $1 - \alpha$ are not necessarily positive, but their sum is equal to 1. This situation is reflected in Fig. 4.3.

5 The Markowitz Model with a Risk-Free Asset

5.1 Data of the Model

Financial Market with a Risk-Free Asset In the previous chapters we considered the Markowitz model of a financial market in which all the assets $i = 1, 2, \ldots, N$ were risky. This was expressed by the assumption that the covariance matrix

$$V = (\sigma_{ij}),\ \sigma_{ij} = Cov(R_i, R_j)$$

of random asset returns was positive definite. Note that this assumption implies

$$\sigma_i^2 = VarR_i = Cov(R_i, R_i) = \langle e_i, Ve_i \rangle > 0,$$

where, according to our standard notation, e_i is the vector with all coordinates equal to 0 except for the ith coordinate which is equal to 1. Thus, in the previous model, the risk (= the variance) associated with each asset return R_i was non-zero.

We now will consider a different model in which one of the assets, $i = 0$, will be risk-free. Its return, $R_0 = r$, will be a strictly positive non-random number. An example of a risk-free asset is cash (holdings in a bank account) with a *fixed* interest rate r. Thus we will deal with a market with $N + 1$ assets indexed by $i = 0, 1, 2, \ldots, N$.

Notation We will use the same notation as before,

$$R = (R_1, \ldots, R_N),$$

for the *N-dimensional vector* with coordinates $R_1, \ldots, R_N$. The $(N + 1)$-*dimensional vector* with coordinates $R_0 = r, R_1, \ldots, R_N$ will be denoted by

I.V. Evstigneev et al., *Mathematical Financial Economics*, Springer Texts in Business and Economics, DOI 10.1007/978-3-319-16571-4_5

using boldface:

$$\mathbf{R} = (R_0, R_1, \ldots, R_N) = (R_0, R).$$

Note that the 0th coordinate $R_0 = r > 0$ of $\mathbf{R}$ is non-random, while the other coordinates $R_1, \ldots, R_N$ are random.

Analogously, portfolios including a risk-free asset will be denoted by

$$\mathbf{x} = (x_0, x_1, \ldots, x_N) = (x_0, x),$$

where x_0 is the amount invested in the risk-free asset (basically, the amount of cash in the portfolio), and $x = (x_1, \ldots, x_N)$ is the portfolio whose positions correspond to risky assets.

The scalar product $\langle \cdot, \cdot \rangle$ is defined in accordance with the previous notation. In particular, we have

$$\langle \mathbf{R}, \mathbf{x} \rangle = \sum_{i=0}^{N} R_i x_i = R_0 x_0 + \sum_{i=1}^{N} R_i x_i = R_0 x_0 + \langle R, x \rangle.$$

Expectations and Covariances As before, we are given the vector $m = (m_1, \ldots, m_N) = (ER_1, \ldots, ER_N)$ of the expected returns of the risky assets $i = 1, 2, \ldots, N$ and the covariance matrix $V = (\sigma_{ij})$, $\sigma_{ij} = Cov(R_i, R_j)$. Additionally, the number $R_0 = r > 0$—the return of the risk-free asset—is given.

We will denote by

$$\mathbf{m} = (r, m_1, \ldots, m_N) = (r, m)$$

the $(N + 1)$-dimensional vector of the expected returns of all the $N + 1$ assets, including the risk-free one.

Normalized and Self-Financing Portfolios In accordance with the previous definitions, a portfolio

$$\mathbf{x} = (x_0, x), \quad \text{where } x = (x_1, \ldots, x_N),$$

is called *self-financing* if

$$x_0 + x_1 + \ldots + x_N = 0,$$

and *normalized* if

$$x_0 + x_1 + \ldots + x_N = 1.$$

Here, we take into account not only the risky positions of the portfolio, but also the risk-free one.

Denote by $\mathbf{e}$ the $(N+1)$-dimensional vector whose coordinates are equal to 1. We can write

$$\mathbf{e} = (1, e),$$

where $e = (1, 1, \ldots, 1)$ is the N-dimensional vector whose coordinates are equal to 1. Consequently,

$$\sum_{i=0}^{N} x_i = \langle \mathbf{e}, \mathbf{x} \rangle = x_0 + \langle e, x \rangle.$$

Thus self-financing portfolios $\mathbf{x}$ satisfy

$$\langle \mathbf{e}, \mathbf{x} \rangle = 0$$

and, for normalized portfolios $\mathbf{x}$, we have

$$\langle \mathbf{e}, \mathbf{x} \rangle = 1.$$

Return on a Portfolio Given a portfolio $\mathbf{x} = (x_0, x_1, \ldots, x_N)$, the expression

$$\mathbf{R}_{\mathbf{x}} = \langle \mathbf{R}, \mathbf{x} \rangle = \sum_{i=0}^{N} R_i x_i = r x_0 + \langle R, x \rangle$$

defines the return on a portfolio $\mathbf{x} = (x_0, x)$ containing the risk-free and N risky assets.

The expected return on a portfolio $\mathbf{x} = (x_0, x_1, \ldots, x_N)$ is given by

$$E\mathbf{R}_{\mathbf{x}} = E\langle \mathbf{R}, \mathbf{x} \rangle = \langle \mathbf{m}, \mathbf{x} \rangle = r x_0 + \langle m, x \rangle \text{ for } \mathbf{x} = (x_0, x), \tag{5.1}$$

where $m = (m_1, \ldots, m_N)$.

The variance of a portfolio return can be computed as follows

$$\begin{aligned} Var\,\mathbf{R}_{\mathbf{x}} &= Var\,\langle \mathbf{R}, \mathbf{x} \rangle \\ &= Var\,[r x_0 + \langle R, x \rangle] = Var\,\langle R, x \rangle \text{ for } \mathbf{x} = (x_0, x) \end{aligned} \tag{5.2}$$

because the variance of a random variable does not change if we add to this random variable any constant. Thus

$$Var\,\mathbf{R}_{\mathbf{x}} = Var\,\langle R, x \rangle = \langle x, Vx \rangle \quad \text{for } \mathbf{x} = (x_0, x),$$

where V is the covariance matrix of the random vector $R = (R_1, \ldots, R_N)$.

We will also use the notation

$$m_{\mathbf{x}} = E\mathbf{R}_{\mathbf{x}}$$

for the expected return and

$$\sigma_{\mathbf{x}}^2 = Var\,\mathbf{R}_{\mathbf{x}}$$

for the variance of the return on the portfolio $\mathbf{x}$.

The assumptions regarding the *risky* assets $i = 1, \ldots, N$ are the same as before in Chaps. 2–4.

Assumption 1 The covariance matrix V is positive definite.

Assumption 2 There are at least two assets i and j $(i, j \geq 1)$ with expected returns $m_i \neq m_j$.

5.2 Portfolio Optimization with a Risk-Free Asset

The Markowitz Problem ($\mathbf{M}_\tau$) The optimization problem of an investor with *risk tolerance* $\tau \geq 0$ in the market with a risk-free asset is as follows:

$$(\mathbf{M}_\tau) \qquad \max_{\mathbf{x} \in R^{N+1}} \{\tau m_{\mathbf{x}} - \sigma_{\mathbf{x}}^2\}$$

subject to

$$\langle \mathbf{e}, \mathbf{x} \rangle = 1.$$

By using the notation $x = (x_1, \ldots, x_N)$ for the vector of the positions of the portfolio $\mathbf{x} = (x_0, x_1, \ldots, x_N)$ corresponding to the risky assets $i = 1, 2, \ldots, N$, we can write the problem ($\mathbf{M}_\tau$) in the following form:

$$(\mathbf{M}_\tau) \qquad \max_{x_0 \in R, x \in R^N} \{\tau(r x_0 + m_x) - \sigma_x^2\}$$

subject to

$$x_0 + \langle e, x \rangle = 1.$$

or, equivalently,

$$(\mathbf{M}_\tau) \qquad \max_{x_0 \in R, x \in R^N} \{\tau(rx_0 + \langle m, x\rangle) - \langle x, Vx\rangle\}$$

subject to

$$x_0 + \langle e, x\rangle = 1.$$

Definition A portfolio $\mathbf{x}^*$ is called (mean-variance) *efficient* if it solves the optimization problem

$$(\mathbf{M}_\tau) \qquad \max_{\mathbf{x} \in R^{N+1}} \{\tau m_{\mathbf{x}} - \sigma_{\mathbf{x}}^2\}$$

subject to

$$\langle \mathbf{e}, \mathbf{x}\rangle = 1$$

for some $\tau \geq 0$.

This definition is fully analogous to the definition of an efficient portfolio in the market where all assets are risky. The following proposition is proved exactly as Proposition 2.1 in Chap. 2 (the only difference is that the dimension N should be replaced by $N + 1$).

Proposition 5.1 *A normalized portfolio* $\mathbf{x}^* \in R^{N+1}$ *is efficient if and only if there exists no normalized portfolio* $\mathbf{x} \in R^{N+1}$ *such that*

$$m_{\mathbf{x}} \geq m_{\mathbf{x}^*} \text{ and } \sigma_{\mathbf{x}}^2 < \sigma_{\mathbf{x}^*}^2.$$

The last two inequalities mean that $\mathbf{x}^*$ solves the optimization problem

$$(\mathbf{M}_\tau) \qquad \min_{\mathbf{x} \in R^{N+1}} \{\sigma_{\mathbf{x}}^2\}$$

subject to

$$m_{\mathbf{x}} \geq \mu$$

and

$$\langle \mathbf{e}, \mathbf{x}\rangle = 1$$

where $\mu = m_{\mathbf{x}^*}$.

5.3 Solution to the Portfolio Selection Problem

Solving Problem ($\mathbf{M}_\tau$) The solution will be given in terms of the inverse of the covariance matrix V, which we will, as before, denote by W:

$$W = V^{-1}.$$

Theorem 5.1 *For each $\tau \geq 0$ the Markowitz portfolio selection problem ($\mathbf{M}_\tau$) has a unique solution $\mathbf{x}_\tau^*$ given by*

$$\mathbf{x}_\tau^* = \mathbf{x}^{MIN} + \frac{\tau}{2}\mathbf{y}^*, \tag{5.3}$$

where

$$\mathbf{x}^{MIN} = (1, 0, \ldots, 0) \; (= \mathbf{x}_0^*) \tag{5.4}$$

and

$$\mathbf{y}^* = (y_0^*, y^*), \tag{5.5}$$

where

$$y_0^* = rC - A, \;\; y^* = W(m - re).$$

We recall the notation used in Theorem 5.1:

$$A = \langle e, Wm \rangle, \;\; B = \langle m, Wm \rangle, \;\; C = \langle e, We \rangle, \;\; D = BC - A^2.$$

Remark 5.1 We can see that the minimum variance portfolio $\mathbf{x}^{MIN}$ is given by the vector $\mathbf{e}_0 = (1, 0, 0, \ldots, 0)$, the $(N + 1)$-dimensional vector whose 0th coordinate is 1 and all the other coordinates are 0. This is not surprising, since the market contains a risk-free asset $i = 0$, and the minimum of the variance $\sigma_\mathbf{x}^2$ (which is zero!) is attained at the portfolio containing only the risk-free asset.

Remark 5.2 Clearly, the portfolio

$$\mathbf{x}^{MIN} = (1, 0, \ldots, 0)$$

is normalized. Therefore the portfolio $\mathbf{y}^*$, involved in the formula

$$\mathbf{x}_\tau^* = \mathbf{x}^{MIN} + \frac{\tau}{2}\mathbf{y}^*,$$

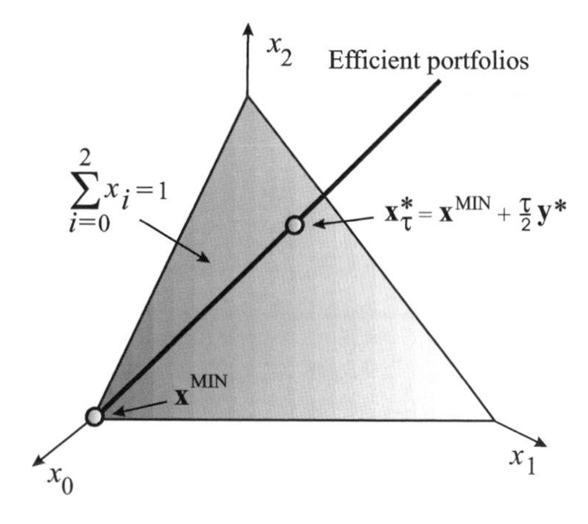

Fig. 5.1 The set of efficient portfolios for the market with a risk-free asset

should be self-financing. As an easy exercise, we can verify this directly:

$$\langle \mathbf{e}, \mathbf{y}^* \rangle = y_0^* + \langle e, y^* \rangle$$
$$= rC - A + \langle e, W(m - re) \rangle = rC - A + A - rC = 0.$$

Remark 5.3 Geometrically, the set of efficient portfolios

$$\mathbf{x}_\tau^* = \mathbf{x}^{MIN} + \frac{\tau}{2}\mathbf{y}^*, \ \tau \geq 0, \tag{5.6}$$

can be represented as a half-line (ray) emanating from the point $\mathbf{x}^{MIN} = (1, 0, \ldots, 0)$ in the $(N + 1)$-dimensional space R^{N+1}. Since all such portfolios are normalized, this line is contained in the hyperplane $x_0 + x_1 + \ldots + x_N = 1$. This situation is illustrated in Fig. 5.1 for $N = 2$ (then there are three coordinates x_0, x_1, x_2).

Proof of Theorem 5.1

1st step. In the Markowitz problem $(\mathbf{M}_\tau)$, we maximize a concave objective function subject to a linear constraint. By virtue of a general result on Lagrange multipliers (see Mathematical Appendix B), a vector $\mathbf{x}_\tau^*$ is a solution to this optimization problem if and only if $\langle \mathbf{e}, \mathbf{x}_\tau^* \rangle = 1$ and there exists a number λ such that $\mathbf{x}_\tau^*$ maximizes the Lagrangian

$$L(\mathbf{x}, \lambda) = \tau m_{\mathbf{x}} - \sigma_{\mathbf{x}}^2 + \lambda(\langle \mathbf{e}, \mathbf{x} \rangle - 1)$$

over all $\mathbf{x} = (x_0, x) \in R^{N+1}$. We have

$$L(\mathbf{x}, \lambda) = \tau(rx_0 + \langle m, x \rangle) - \langle x, Vx \rangle + \lambda(x_0 + \langle e, x \rangle - 1).$$

This function is concave and differentiable in $\mathbf{x} = (x_0, x)$, and so it attains its global maximum at a point at which its partial derivatives with respect to x_0 and x_i $(i = 1, \ldots, N)$ are equal to zero.

By differentiating

$$L(\mathbf{x}, \lambda) = \tau(r x_0 + \langle m, x \rangle) - \langle x, Vx \rangle + \lambda(x_0 + \langle e, x \rangle - 1)$$

with respect to x_0, we find

$$\tau r + \lambda = 0.$$

By computing the vector of partial derivatives (the gradient) of $L(\mathbf{x}, \lambda)$ with respect to $x = (x_1, \ldots, x_N)$, we obtain

$$\tau m - 2Vx + \lambda e = 0.$$

This follows from the facts which we have already used: the gradient of $\langle m, x \rangle$ is m, and the gradient of $\langle x, Vx \rangle$ is $2Vx$.

Thus we have

$$\tau r + \lambda = 0. \tag{5.7}$$

and

$$\tau m - 2Vx + \lambda e = 0. \tag{5.8}$$

The constraint $\langle \mathbf{e}, \mathbf{x} \rangle = 1$ can be written

$$x_0 + \langle e, x \rangle = 1. \tag{5.9}$$

A vector $\mathbf{x} = (x_0, x)$ is a solution to $(\mathbf{M}_\tau)$ if and only if (x_0, x, λ) is a solution to system (5.7)–(5.9) for some λ.

2nd step. Let us solve the above system of equations. We will show that a solution (x_0, x, λ) exists and is unique. This will prove the existence and uniqueness of a solution to $(\mathbf{M}_\tau)$.

From (5.7), $\lambda = -\tau r$. By substituting λ into (5.8), we get

$$\tau m - 2Vx - \tau r e = 0,$$

which yields $Vx = \frac{\tau}{2}(m - re)$. Since $W = V^{-1}$, we find

$$x = \frac{\tau}{2} W(m - re).$$

In view of (5.9), and since $A = \langle e, Wm \rangle$, $C = \langle e, We \rangle$, we get

$$x_0 = 1 - \langle e, x \rangle = 1 - \frac{\tau}{2}\langle e, W(m - re)\rangle = 1 - \frac{\tau}{2}(A - rC).$$

3rd step. We have obtained that the problem $(\mathbf{M}_\tau)$ has the following unique solution

$$\mathbf{x}_\tau^* = (1 - \frac{\tau}{2}(A - rC), \frac{\tau}{2}W(m - re)).$$

This vector is equal to the sum of the vector

$$\mathbf{x}^{MIN} = (1, 0, 0, \ldots, 0)$$

and the vector $\frac{\tau}{2}\mathbf{y}^*$, where

$$\mathbf{y}^* = (rC - A, W(m - re)).$$

This completes the proof.

□

6 Efficient Portfolios in a Market with a Risk-Free Asset

6.1 Expectations and Variances of Portfolio Returns

The expected return of the portfolio $\mathbf{x}_\tau^*$ is equal to

$$
\begin{aligned}
m_{\mathbf{x}_\tau^*} = \langle \mathbf{m}, \mathbf{x}_\tau^* \rangle &= r[1 - \frac{\tau}{2}(A - rC)] + \langle m, \frac{\tau}{2} W(m - re) \rangle \\
&= r + \frac{\tau}{2}[r^2 C - rA + \langle m, Wm \rangle - r \langle m, We \rangle] \\
&= r + \frac{\tau}{2}[r^2 C - rA + B - rA] \\
&= r + \frac{\tau}{2}[r^2 C - 2rA + B],
\end{aligned}
$$

where, we recall,

$$
\begin{aligned}
A &= \langle e, Wm \rangle = \langle We, m \rangle = \langle m, We \rangle, \\
B &= \langle m, Wm \rangle, \; C = \langle e, We \rangle.
\end{aligned}
$$

Let us introduce the following notation:

$$H = r^2 C - 2rA + B.$$

Then we can write:

$$m_{\mathbf{x}_\tau^*} = E\,\mathbf{R}_{\mathbf{x}_\tau^*} = r + \frac{\tau}{2} H.$$

The variance of the return of the portfolio $\mathbf{x}_\tau^*$. Recall the general formula

$$Var\,\mathbf{R}_\mathbf{x} = Var\,\langle R, x \rangle = \langle x, Vx \rangle \text{ for } \mathbf{x} = (x_0, x).$$

I.V. Evstigneev et al., *Mathematical Financial Economics*, Springer Texts in Business and Economics, DOI 10.1007/978-3-319-16571-4_6

By applying this formula to the portfolio

$$\mathbf{x}_\tau^* = \mathbf{x}^{MIN} + \frac{\tau}{2}\mathbf{y}^* = (1 + \frac{\tau}{2}(rC - A),\ \frac{\tau}{2}W(m - re)),$$

we get

$$Var\,\mathbf{R}_{\mathbf{x}_\tau^*} = \langle \frac{\tau}{2}W(m - re), V\frac{\tau}{2}W(m - re)\rangle = \frac{\tau^2}{4}\langle W(m - re), (m - re)\rangle.$$

Here,

$$\begin{aligned}\langle W(m - re), (m - re)\rangle &= \langle Wm, m\rangle - r\langle We, m\rangle - r\langle Wm, e\rangle + r^2\langle We, e\rangle \\ &= r^2C - 2rA + B = H.\end{aligned}$$

Thus

$$\sigma^2_{\mathbf{x}_\tau^*} = Var\,\mathbf{R}_{\mathbf{x}_\tau^*} = \frac{\tau^2}{4}H.$$

We have obtained the following **formulas for the expected return and the variance of the return for the efficient portfolio $\mathbf{x}_\tau^*$:**

$$m_{\mathbf{x}_\tau^*} = r + \frac{\tau}{2}H, \tag{6.1}$$

$$\sigma^2_{\mathbf{x}_\tau^*} = \frac{\tau^2}{4}H, \tag{6.2}$$

where $H = r^2C - 2rA + B$.

Let us show that $H > 0$. Indeed, the minimum value of the quadratic function $\phi(r) = r^2C - 2rA + B$ is attained at $r = A/C$, and this value is

$$\frac{A^2}{C^2}C - 2\frac{A}{C}A + B = \frac{BC - A^2}{C} = \frac{D}{C},$$

where, as we showed in the previous chapters, $D > 0$ and $C > 0$.

6.2 Efficient Frontier and the Capital Market Line

Efficient Frontier The results obtained enable us to give a complete description of the *efficient frontier* for the market with a risk-free asset. As before, the efficient frontier is defined as the set of those points in the plane which correspond to the variances and the expectations of returns on efficient portfolios. (Recall that efficient

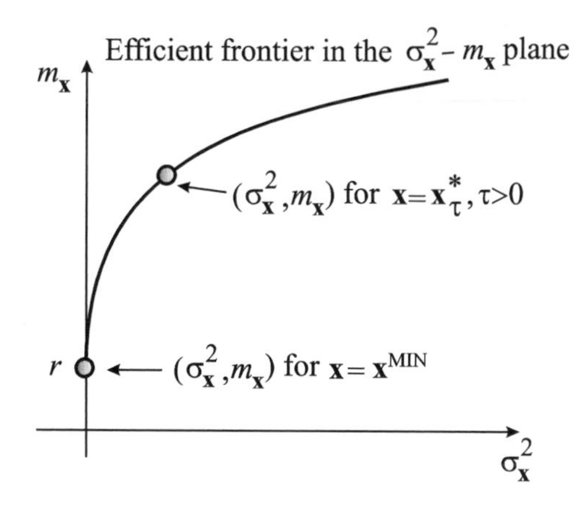

Fig. 6.1 Market with a risk-free asset: The efficient frontier in the $\sigma_{\mathbf{x}}^2$-$m_{\mathbf{x}}$ plane

portfolios, by definition, are solutions to the Markowitz problem $(\mathbf{M}_\tau)$, where τ ranges through all non-negative numbers.) Thus in the plane with coordinates $\sigma_{\mathbf{x}}^2, m_{\mathbf{x}}$, the efficient frontier is the curve consisting of the points

$$\sigma_{\mathbf{x}}^2 = \frac{\tau^2}{4}H, \; m_{\mathbf{x}} = r + \frac{\tau}{2}H, \; \tau \geq 0.$$

Excluding τ from these equations, we obtain that the curve under consideration is (the upper portion of) the *parabola* with vertex at $(\sigma_{\mathbf{x}}^2, m_{\mathbf{x}}) = (0, r)$:

$$\sigma_{\mathbf{x}}^2 = \frac{(m_{\mathbf{x}} - r)^2}{H}.$$

The efficient frontier in the $\sigma_{\mathbf{x}}^2$-$m_{\mathbf{x}}$ plane is depicted in Fig. 6.1.

Capital Market Line We can also draw the efficient frontier in the $\sigma_{\mathbf{x}}$-$m_{\mathbf{x}}$ plane, where the coordinates are the standard deviations $\sigma_{\mathbf{x}}$ and the expectations $m_{\mathbf{x}}$ of returns. We have

$$\sigma_{\mathbf{x}}^2 = \frac{(m_{\mathbf{x}} - r)^2}{H},$$

and so

$$\sigma_{\mathbf{x}} = \frac{m_{\mathbf{x}} - r}{\sqrt{H}}.$$

The set of pairs $(\sigma_{\mathbf{x}}, m_{\mathbf{x}})$ satisfying this equation is the *straight line* emanating from the point $(\sigma_{\mathbf{x}}, m_{\mathbf{x}}) = (0, r)$, see the diagram below. This line is called the *capital market line*. It is shown in the Fig. 6.2.

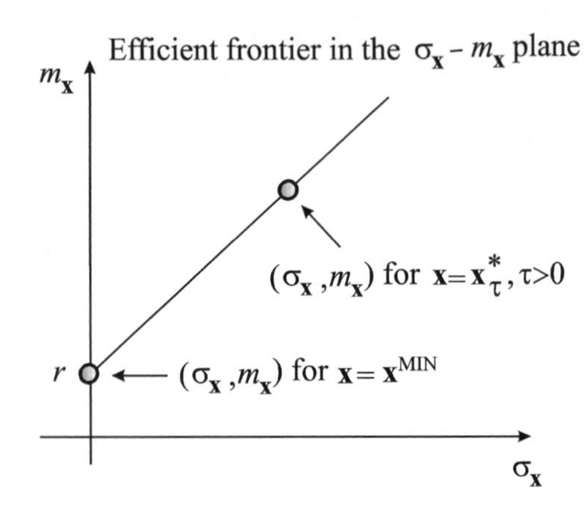

Fig. 6.2 The capital market line

6.3 Tangency Portfolio

Definition of the Tangency Portfolio We have shown that all efficient portfolios in a market with a risk-free asset are described by the formula

$$\mathbf{x}^*_\tau = \mathbf{x}^{MIN} + \frac{\tau}{2}\mathbf{y}^*, \ \ \tau \geq 0,$$

where $\mathbf{x}^{MIN} = (1, 0, 0, \ldots, 0)$ and $\mathbf{y}^* = (rC - A, W(m - re))$. We can also write

$$\mathbf{x}^*_\tau = (1 + \frac{\tau}{2}(rC - A), \frac{\tau}{2}W(m - re)). \tag{6.3}$$

Consider that value of τ for which the 0th coordinate of the vector $\mathbf{x}^*_\tau$ is equal to 0:

$$1 + \frac{\tau}{2}(rC - A) = 0.$$

We can see that this is true for $\tau = \bar{\tau}$, where

$$\bar{\tau} = \frac{2}{A - rC}.$$

This number is positive if and only if the following assumption holds (which we will impose in the current context):

Assumption 3

$$r < \frac{A}{C}.$$

Under this assumption, the portfolio $\mathbf{v} = \mathbf{x}^*_{\bar{\tau}}$ is efficient; it will be termed the tangency portfolio. Thus we give the following definition.

Definition The portfolio $\mathbf{v} = \mathbf{x}^*_{\bar{\tau}}$, where $\bar{\tau} = \dfrac{2}{A - rC}$ (> 0), is called the *tangency portfolio*.

Why the term "tangency" is used here will be explained later.

Remark Recall that A/C is the expected return on the minimum variance portfolio in the market with N risky assets $i = 1, 2, \ldots, N$ whose random returns are $R_1, \ldots, R_N$. Assumption 3, according to which $r < \dfrac{A}{C}$, means that the return on the risk-free asset is less than the expected return on every efficient portfolio containing only risky assets. Thus, although assets $i = 1, 2, \ldots, N$ are risky, there are substantial incentives to include them in a portfolio, rather than dealing only with risk-free but relatively small returns from asset $i = 0$.

Formula for the Tangency Portfolio By definition, we have $\mathbf{v} = \mathbf{x}^*_{\bar{\tau}}$, where $\bar{\tau} = 2/(A - rC)$. By setting $\tau = \bar{\tau}$ in the formula (6.3) for an efficient portfolio, we get

$$\mathbf{v} = \left(0, \frac{W(m - re)}{A - rC}\right),$$

or,

$$\mathbf{v} = (0, v),$$

where

$$v = \frac{W(m - re)}{A - rC}.$$

Efficient Frontiers and the Tangency Portfolio By construction, the tangency portfolio $\mathbf{v} = \mathbf{x}^*_{\bar{\tau}}$ *contains only risky assets*: its 0th position is equal to 0. Thus, $\mathbf{v} = (0, v)$, where v is a portfolio of N risky assets. Since $\mathbf{v} = \mathbf{x}^*_{\bar{\tau}}$, the portfolio $\mathbf{v}$ is the solution to the Markowitz problem $(\mathbf{M}_{\bar{\tau}})$, i.e., $\mathbf{v} = (0, v)$ maximizes the objective function

$$\bar{\tau} m_{\mathbf{x}} - \sigma^2_{\mathbf{x}} = \bar{\tau}(r x_0 + \langle m, x \rangle) - \langle x, Vx \rangle$$

over all portfolios $\mathbf{x} = (x_0, x)$ satisfying the normalization constraint $\langle \mathbf{e}, \mathbf{x} \rangle = 1$, which can be written as

$$x_0 + \langle e, x \rangle = 1.$$

In particular, the value of the above objective function for the portfolio $\mathbf{v} = (0, v)$ is not less than its value for each normalized portfolio of the form $(0, x)$, where $x \in R^N$. Thus

$$\bar{\tau}\langle m, v\rangle - \langle v, Vv\rangle \geq \bar{\tau}\langle m, x\rangle - \langle x, Vx\rangle$$

for each $x \in R^N$ satisfying $\langle e, x\rangle = 1$. Therefore

$$v = x^*_{\bar{\tau}}$$

i.e., v is nothing but the portfolio $x^*_{\bar{\tau}}$, the solution to problem $(\mathbf{M}_{\bar{\tau}})$ considered *for the market with N risky assets.*

Furthermore, since $\mathbf{v} = (0, v) = (0, x^*_{\bar{\tau}})$, we have

$$(\sigma_{\mathbf{v}}, m_{\mathbf{v}}) = (\sigma_v, m_v) = (\sigma_{x^*_{\bar{\tau}}}, m_{x^*_{\bar{\tau}}}).$$

Thus *the point (σ_v, m_v) belongs both to the efficient frontier of the market of $N + 1$ assets and the efficient frontier of the market of N assets.*

In the efficient frontier of the market of $N + 1$ assets with vector of returns $\mathbf{R} = (r, R_1, \ldots, R_N)$, the point (σ_v, m_v) corresponds to the efficient portfolio $\mathbf{x}^*_{\bar{\tau}} \in R^{N+1}$ with risk tolerance $\bar{\tau}$.

In the efficient frontier of the market of N assets with vector of returns $R = (R_1, \ldots, R_N)$, the point (σ_v, m_v) corresponds to the efficient portfolio $x^*_{\bar{\tau}} \in R^N$ with (the same) risk tolerance $\bar{\tau}$.

Recall that the efficient frontier (in the σ_x-m_x plane) for the market of N risky assets is the hyperbola:

$$\sigma_x = \sqrt{\frac{1}{D}(Cm_x^2 - 2Am_x + B)}$$

with vertex at $(\sqrt{1/C}, A/C)$. The efficient frontier for the market with a risk-free asset is the straight line

$$\sigma_{\mathbf{x}} = \frac{m_{\mathbf{x}} - r}{\sqrt{H}}.$$

The tangency portfolio corresponds to *the point where the straight line intersects the hyperbola*. It can be shown that this point is unique and, at this point, the straight line is *tangent* to the hyperbola. This property motivates the term "tangency portfolio." Figure 6.3 gives an illustration.

The Sharpe Ratio With each portfolio $\mathbf{x} \in R^{N+1}$, we associate the point in the plane having the coordinates

$$\sigma = \sigma_{\mathbf{x}}, \ \mu = m_{\mathbf{x}}.$$

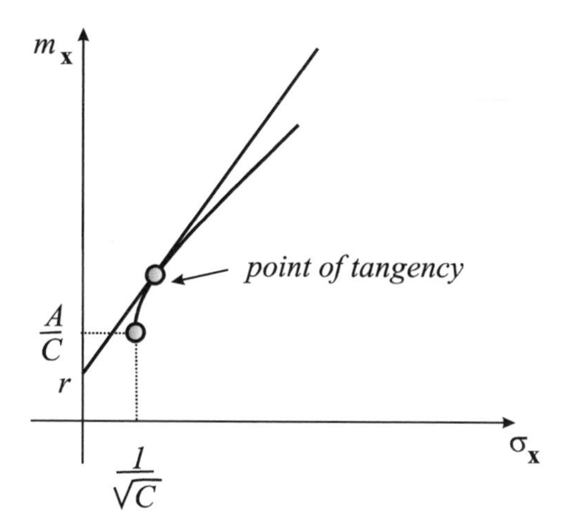

Fig. 6.3 The tangency portfolio: A geometric illustration

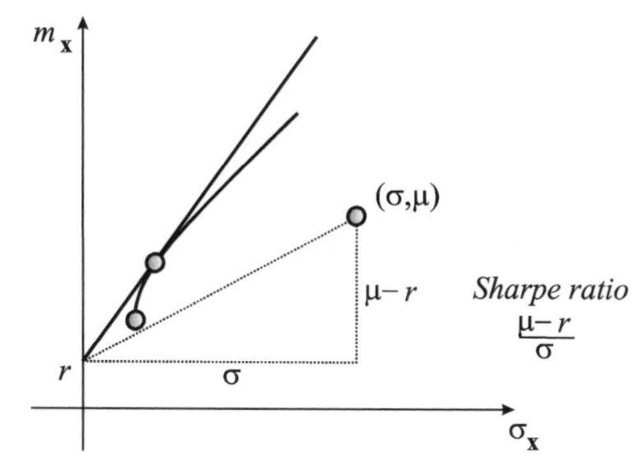

Fig. 6.4 Sharpe ratio

If $\sigma > 0$, we define

$$\rho_{\mathbf{x}} = \frac{\mu - r}{\sigma}.$$

The number $\rho_{\mathbf{x}}$ is called the *Sharpe ratio* of the portfolio $\mathbf{x}$. The ratio $\rho_{\mathbf{x}}$ is *the expectation of excess return per unit risk*. Here, risk is estimated in terms of the standard deviation $\sigma = \sigma_{\mathbf{x}}$ of the portfolio return. The expectation of excess return is given by $\mu - r = E\mathbf{R}_{\mathbf{x}} - r$. We can see from the diagram below that the tangency portfolio has the greatest Sharpe ratio among all efficient portfolios without risk-free asset. The Sharpe ratio is the same for all efficient portfolios with risk-free asset (Fig. 6.4).

6.4 A Mutual Fund Theorem

Tobin's Mutual Fund Theorem By virtue of Theorem 5.1, every efficient portfolio in the market with a risk-free asset can be represented as

$$\mathbf{x}^*_\tau = \mathbf{x}^{MIN} + \frac{\tau}{2}\mathbf{y}^*, \ \tau \geq 0,$$

where $\mathbf{x}^{MIN} = (1, 0, \ldots, 0)$. From this we obtain the following result:

Theorem 6.1 *Every efficient portfolio* $\mathbf{x}$ *can be represented as a combination*

$$\mathbf{x} = \alpha\mathbf{x}^{MIN} + (1 - \alpha)\mathbf{v}$$

of the risk-free portfolio $\mathbf{x}^{MIN}$ *and the tangency portfolio* $\mathbf{v} = \mathbf{x}^*_{\bar{\tau}}$.

Proof of Theorem 6.1 Let us represent $\mathbf{x}$ in the form $\mathbf{x} = \mathbf{x}^{MIN} + \frac{\tau}{2}\mathbf{y}^*$, where τ is a non-negative number. Define

$$\alpha = \frac{\bar{\tau} - \tau}{\bar{\tau}}.$$

Then, as is easily seen, $(1 - \alpha)\bar{\tau} = \tau$, and so

$$\begin{aligned}\alpha\mathbf{x}^{MIN} + (1 - \alpha)\mathbf{v} &= \alpha\mathbf{x}^{MIN} + (1 - \alpha)[\mathbf{x}^{MIN} + \frac{\bar{\tau}}{2}\mathbf{y}^*] \\ &= \mathbf{x}^{MIN} + \frac{\tau}{2}\mathbf{y}^* = \mathbf{x},\end{aligned}$$

which completes the proof. □

The above result is known as *Tobin's*[1] *mutual-fund theorem.* It states that any efficient portfolio is a combination of a risk-free asset and a "mutual fund": the efficient portfolio $\mathbf{v}$ (tangency portfolio) consisting of only risky assets.

A Geometric Illustration We have seen that all the vectors representing efficient portfolios form a half-line (ray) in the $(N + 1)$-dimensional space. Geometrically, the assertion of Theorem 6.1 can be illustrated as follows. Suppose that we fix two points, corresponding to the minimum variance portfolio $\mathbf{x}^{MIN}$ and the tangency portfolio $\mathbf{v}$, on this ray. Then any other point (efficient portfolio $\mathbf{x}$) on this ray can be obtained as a weighted average

[1] James Tobin, 1981 Nobel Laureate in Economics.

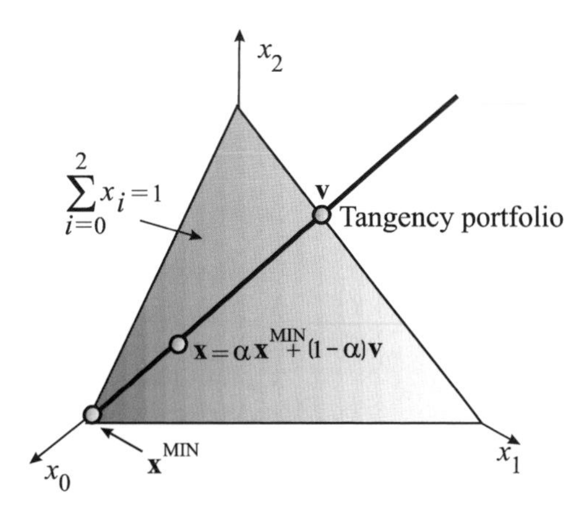

Fig. 6.5 Tobin's theorem: A geometric illustration

$$\mathbf{x} = \alpha \mathbf{x}^{MIN} + (1 - \alpha)\mathbf{v}$$

of these two points. The "weight" α can be any number not greater than 1. Figure 6.5 provides an illustration.

Capital Asset Pricing Model (CAPM) 7

7.1 A General Result

An Introduction to the CAPM "CAPM" means "Capital Asset Pricing Model," although this term is often associated with a certain *equation* (rather than a model), a general version of which we are going to derive now.

We begin with a general observation. Consider the Markowitz model for a financial market containing N risky assets $i = 1, 2, \ldots, N$, and one risk-free asset, $i = 0$. The returns R_i on assets $i = 1, 2, \ldots, N$ are random, and the return $R_0 = r > 0$ on asset 0 is constant. As before, we denote by V the covariance matrix of the random vector $R = (R_1, \ldots, R_N)$ and by $m = (m_1, \ldots, m_N)$ the vector of expected returns $m_i = ER_i$. Assumptions 1 and 2 (m and e are not collinear, and V is positive definite), see Chap. 2, are supposed to hold.

Let

$$\mathbf{x} = (x_0, x), \ x = (x_1, \ldots, x_N),$$

$$\mathbf{y} = (y_0, y), \ y = (y_1, \ldots, y_N)$$

be two portfolios. Consider the random variables describing the returns on the portfolios $\mathbf{x}$ and $\mathbf{y}$: $\mathbf{R_x} = R_0 x_0 + \sum_{i=1}^N R_i x_i$ and $\mathbf{R_y} = R_0 y_0 + \sum_{i=1}^N R_i y_i$. Here $R_0 x_0$ and $R_0 y_0$ are (non-random) constants. Therefore

$$Cov(\mathbf{R_x}, \mathbf{R_y}) = Cov(R_x, R_y) = \langle x, Vy \rangle, \tag{7.1}$$

[see (4.1)].

Let us consider some efficient portfolio $\mathbf{x} = \mathbf{x}_\tau^*$ with $\tau > 0$ in the Markowitz model under consideration. Recall that efficient portfolios are solutions to the portfolio selection problem $(\mathbf{M}_\tau)$, where τ (risk tolerance) is any non-negative number.

I.V. Evstigneev et al., *Mathematical Financial Economics*, Springer Texts in Business and Economics, DOI 10.1007/978-3-319-16571-4_7

Theorem 7.1 *For each* $i = 1, 2, \ldots, N$, *the following equation holds:*

$$ER_i - r = \frac{Cov(R_i, \mathbf{R_x})}{Var\mathbf{R_x}}(E\mathbf{R_x} - r). \tag{7.2}$$

Formula (7.2) is the (general version of) CAPM.

Proof of Theorem 7.1

1st step. We have proved that

$$E\mathbf{R}_{\mathbf{x}_\tau^*} = r + \frac{\tau}{2}H, Var\mathbf{R}_{\mathbf{x}_\tau^*} = \frac{\tau^2}{4}H$$

where $H > 0$ is some number ($H = r^2C - 2rA + B$). Consequently,

$$\frac{E\mathbf{R}_{\mathbf{x}_\tau^*} - r}{Var\mathbf{R}_{\mathbf{x}_\tau^*}} = \frac{2}{\tau}, \ \tau > 0.$$

2nd step. For each $i = 1, 2, \ldots, N$, let us compute the covariance of the random variables R_i and $\mathbf{R}_{\mathbf{x}_\tau^*}$. Note that R_i is the return on the portfolio $\mathbf{e}_i = (0, e_i)$, where $e_i = (0, \ldots, 1, \ldots 0)$ (the ith coordinate equals 1). Also, recall that

$$\mathbf{x}_\tau^* = (1 + \frac{\tau}{2}(rC - A), \frac{\tau}{2}W(m - re)).$$

Consequently,

$$Cov(R_i, \mathbf{R}_{\mathbf{x}_\tau^*}) = \langle e_i, V\frac{\tau}{2}W(m - re)\rangle$$

(see formula (7.1) above). Thus we obtain

$$Cov(R_i, \mathbf{R}_{\mathbf{x}_\tau^*}) = \langle e_i, V\frac{\tau}{2}W(m - re)\rangle$$
$$= \langle e_i, \frac{\tau}{2}(m - re)\rangle = \frac{\tau}{2}(m_i - r)$$

(we have used that $W = V^{-1}$).

3rd step. By combining the result obtained at the 2nd step,

$$m_i - r = \frac{2}{\tau}Cov(R_i, \mathbf{R}_{\mathbf{x}_\tau^*}),$$

and the result obtained at the 1st step,

$$\frac{2}{\tau} = \frac{E\mathbf{R}_{\mathbf{x}_\tau^*} - r}{Var\mathbf{R}_{\mathbf{x}_\tau^*}},$$

we conclude

$$m_i - r = \frac{Cov(R_i, \mathbf{R}_{\mathbf{x}_\tau^*})}{Var\mathbf{R}_{\mathbf{x}_\tau^*}}(E\mathbf{R}_{\mathbf{x}_\tau^*} - r),$$

which is precisely the formula (7.2) with $\mathbf{x}_\tau^* = \mathbf{x}$.

The proof is complete. □

7.2 An Equilibrium Approach to the CAPM

Net Return on a Portfolio of Value w Suppose an investor having positive wealth $w > 0$ wishes to invest it in assets $i = 0, 1, \ldots, N$, i.e., to select a portfolio $\mathbf{x} = (x_0, x_1, \ldots, x_N)$ with

$$x_0 + x_1 + \ldots + x_N = w.$$

We know (see Chap. 1) that the net return on the portfolio $\mathbf{x}$, i.e.,

$$\frac{(\text{value of } \mathbf{x} \text{ at time } 1) - (\text{value of } \mathbf{x} \text{ at time } 0)}{(\text{value of } \mathbf{x} \text{ at time } 0)},$$

can be computed as follows:

$$\sum_{i=0}^{N} x_i R_i / \sum x_i = \frac{1}{w} \sum_{i=0}^{N} x_i R_i = \sum_{i=0}^{N} \frac{x_i}{w} R_i = \mathbf{R}_{\mathbf{x}/w}.$$

We can see from these relations that the net return on the portfolio $\mathbf{x}$ coincides with the return on the normalized portfolio

$$\mathbf{x}/w = (\frac{x_0}{w}, \frac{x_1}{w}, \ldots, \frac{x_N}{w}),$$

so that

$$\frac{\mathbf{R}_\mathbf{x}}{w} = \mathbf{R}_{\mathbf{x}/w}. \tag{7.3}$$

Efficient Portfolios of Initial Value *w* In the Markowitz model, an investor having initial wealth w wants to maximize the expectation of net return and minimize its variance. Formally, the optimization problem for an investor with wealth $w > 0$ is given by:

$$(\mathbf{M}_\tau(w)) \qquad \max_{\mathbf{x}\in R^{N+1}} \{\tau E(\frac{\mathbf{R_x}}{w}) - Var(\frac{\mathbf{R_x}}{w})\}$$

subject to

$$\langle \mathbf{e}, \mathbf{x} \rangle = w.$$

Recall that $\mathbf{e} = (1, 1, \ldots, 1) \in R^{N+1}$, and so $\langle \mathbf{e}, \mathbf{x} \rangle = x_0 + \ldots + x_N$. Here, as before, the number $\tau \geq 0$ stands for the risk tolerance of the investor. Solutions to $(\mathbf{M}_\tau(w))$, where τ ranges through non-negative numbers, are *efficient portfolios of initial value* w.

How can we solve problem $(\mathbf{M}_\tau(w))$? Quite easily, as long as we know the solution to $(\mathbf{M}_\tau)$. Problem $(\mathbf{M}_\tau(w))$ reduces to $(\mathbf{M}_\tau)$ (for which $w = 1$) as follows. We have

$$\langle \mathbf{e}, \mathbf{x} \rangle = w \text{ if and only if } \langle \mathbf{e}, \mathbf{x}/w \rangle = 1,$$

and

$$\tau E(\frac{\mathbf{R_x}}{w}) - Var(\frac{\mathbf{R_x}}{w}) = \tau E(\mathbf{R}_{\mathbf{x}/w}) - Var(\mathbf{R}_{\mathbf{x}/w})$$

because $\mathbf{R_x}/w = \mathbf{R}_{\mathbf{x}/w}$ [see (7.3)]. Thus $\mathbf{x}$ is a solution to $(\mathbf{M}_\tau(w))$ if and only if $\mathbf{x}/w$ is a solution to $(\mathbf{M}_\tau)$. Consequently, problem $(\mathbf{M}_\tau(w))$ has the following unique solution

$$\mathbf{x}^*_\tau(w) = w\mathbf{x}^*_\tau,$$

where the portfolio $\mathbf{x}^*_\tau$ is the solution to the Markowitz problem $(\mathbf{M}_\tau)$.

CAPM: An Equilibrium Model Now we are going to derive some special version of the CAPM in which $\mathbf{x}$ will be the so-called market portfolio (to be defined below). To this end we will outline an equilibrium model for the financial market under consideration. Recall that in this market there is a risk-free asset ($i = 0$) with return r and N risky assets $i = 1, 2, \ldots, N$,whose returns are random variables

R_i.Suppose that there are K investors $k = 1, 2, \ldots, K$ trading in the market. Let the following assumptions hold:

(a) Investors $k = 1, 2, \ldots, K$ have full information about the expectations and the covariances of the random returns R_i, $i = 1, 2, \ldots, N$.
(b) The vectors $e = (1, 1, \ldots, 1)$ and $m = (ER_1, \ldots, ER_N)$ are not collinear, and the matrix $V = (Cov(R_i, R_j))$ is positive definite (i.e. the standard assumptions 1 and 2 are supposed to hold).
(c) Each investor k possesses an initial wealth $w^{(k)} > 0$ and invests it all in the assets $i = 0, 1, 2, \ldots, N$.
(d) Each investor k has a risk tolerance $\tau_k > 0$ and chooses the corresponding mean-variance efficient portfolio.
(e) The market is in equilibrium:

$$\text{asset supply} = \text{asset demand}.$$

The assumption (e) means that every asset is in the portfolio of some investor.

Investors' Portfolios Let

$$\mathbf{x}^{(k)} = (x_0^{(k)}, \ldots, x_N^{(k)})$$

be the portfolio of investor k, i.e., let $x_i^{(k)}$ be the amount of money invested in asset i by investor k at time 0 (at the beginning of the investment period). According to assumption (c), we have

$$x_0^{(k)} + \ldots + x_N^{(k)} = w^{(k)}.$$

By virtue of assumption (d), each investor selects the efficient portfolio with initial value $w^{(k)}$. As we have shown, this portfolio is as follows:

$$\mathbf{x}^{(k)} = w^{(k)} \, \mathbf{x}^*_{\tau_k},$$

where τ_k is the investor's risk tolerance. According to our notation, $\mathbf{x}^*_\tau$ is the solution to the Markowitz problem $(\mathbf{M}_\tau)$. By virtue of the results obtained in Theorem 5.1, $\mathbf{x}^*_\tau = \mathbf{x}^{MIN} + \frac{\tau}{2}\mathbf{y}^*$, where $\mathbf{x}^{MIN}$ and $\mathbf{y}^*$ are given by (5.4) and (5.5).

Market Portfolio Observe that total wealth invested in asset i by all the investors is

$$w_i = \sum_{k=1}^{K} x_i^{(k)}.$$

According to the equilibrium condition (e) (supply is equal to demand), w_i is the cost of all the assets of type i traded in the market. Total market wealth w is equal to the sum

$$w = w_0 + \ldots + w_N.$$

On the other hand,

$$w = w^{(1)} + \ldots + w^{(K)}$$

because

$$\sum_{i=0}^{N} w_i = \sum_{i=0}^{N} \sum_{k=1}^{K} x_i^{(k)} = \sum_{k=1}^{K} \sum_{i=0}^{N} x_i^{(k)} = \sum_{k=1}^{K} w^{(k)}.$$

Definition The portfolio $\mathbf{x}^M = (x_0^M, \ldots, x_N^M)$, where

$$x_i^M = \frac{w_i}{w}, \; i = 0, 1, 2, \ldots, N,$$

is called *the market portfolio*.

This is a normalized portfolio (since $\sum \frac{w_i}{w} = 1$), and its positions represent the proportions of market wealth invested in assets $i = 0, 1, 2, \ldots, N$ (*capitalization weights*). Since $w_i = \sum_k x_i^{(k)}$, we have

$$\mathbf{x}^M = \frac{\mathbf{x}^{(1)} + \ldots + \mathbf{x}^{(K)}}{w}.$$

The Market Portfolio Is Efficient This immediately follows from the relations

$$\mathbf{x}^M = \frac{1}{w} \sum_{k=1}^{K} \mathbf{x}^{(k)} = \frac{1}{w} \sum_{k=1}^{K} w^{(k)} \mathbf{x}_{\tau_k}^*$$

$$= \frac{1}{w} \sum_{k=1}^{K} w^{(k)} (\mathbf{x}^{MIN} + \frac{\tau_k}{2} \mathbf{y}^*)$$

$$= \mathbf{x}^{MIN} + \frac{1}{w} \sum_{k=1}^{K} w^{(k)} \frac{\tau_k}{2} \mathbf{y}^* = \mathbf{x}^{MIN} + \frac{\tau}{2} \mathbf{y}^*,$$

where

$$\tau = \frac{1}{w}\sum_{k=1}^{K} w^{(k)}\tau_k.$$

These relations imply that the market portfolio is an efficient normalized portfolio corresponding to the level of risk tolerance τ, a weighted average of the risk tolerance coefficients τ_k of investors $k = 1, 2, \ldots, K$. For each k, the weight $w^{(k)}/w$ reflects relative investor k's wealth.

7.3 The Sharpe-Lintner-Mossin Formula

CAPM and the Market Portfolio By virtue of Theorem 7.1, formula (7.2) holds for any efficient portfolio $\mathbf{x}$. Since the market portfolio $\mathbf{x}^M$ is efficient, we can substitute $\mathbf{x} = \mathbf{x}^M$ into this formula. To alleviate notation, we will write $\mathbf{R}^M = \mathbf{R}_{\mathbf{x}^M}$ for the return on the market portfolio. Then we arrive at the equation

$$ER_i - r = \frac{Cov(R_i, \mathbf{R}^M)}{Var\mathbf{R}^M}(E\mathbf{R}^M - r),\ i = 1, 2, \ldots, N.$$

This is the celebrated Sharpe-Lintner-Mossin CAPM equation.[1]

The Beta of an Asset The coefficient

$$\beta_i = \frac{Cov(R_i, \mathbf{R}^M)}{Var\mathbf{R}^M}$$

involved in the CAPM formula is called the *beta coefficient of asset* i. The difference $ER_i - r$ is the *risk premium* on asset i. The difference $E\mathbf{R}^M - r$ is the *risk premium* on the market portfolio.

By using the notation $m_i = ER_i$ and $m^M = E\mathbf{R}^M$, we can briefly write the CAPM equation as follows:

$$m_i - r = \beta_i(m^M - r).$$

The CAPM relation shows that the number $\beta_i = Cov(R_i, \mathbf{R}^M)/Var\mathbf{R}^M$, rather than $VarR_i$, is crucial for the evaluation of the risk premium on asset i. This observation has led to an important conceptual change in economic modelling.

[1] William F. Sharpe was honored in 1990 with the Nobel Prize in Economics.

CAPM Continued

8

8.1 Security Market Line and the Pricing Formula

Security Market Line The set of those points in the β_i-ER_i plane which satisfy

$$ER_i = r + \beta_i(E\mathbf{R}^M - r)$$

is called the *security market line*. The security market line reflects a linear dependence of the expected return of asset i on its beta coefficient. The slope of this line is equal to the risk premium on the market portfolio.

The following Fig. 8.1 illustrates the security market line.

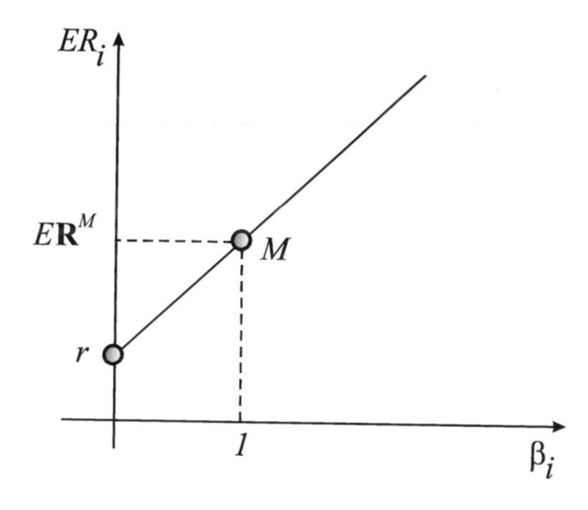

Fig. 8.1 Security market line

I.V. Evstigneev et al., *Mathematical Financial Economics*, Springer Texts in Business and Economics, DOI 10.1007/978-3-319-16571-4_8

The point M corresponds to the market portfolio: if $\beta = 1$, then $r + \beta(E\mathbf{R}^M - r) = E\mathbf{R}^M$.

CAPM as a Pricing Formula In the standard formulation of the CAPM, prices are not contained explicitly. Only the expected rates of return appear in the formula. To understand why the CAPM is interpreted as a pricing model, let us recall the definition of return:

$$R_i = \frac{S_1^i - S_0^i}{S_0^i},$$

where S_0^i and S_1^i are the prices of asset i at the beginning and at the end of the investment period (at time 0 and at time 1). Here S_1^i is random, while S_0^i is deterministic. By substituting R_i into the CAPM formula, we find

$$\frac{ES_1^i - S_0^i}{S_0^i} = r + \beta_i(E\mathbf{R}^M - r),$$

which gives

$$S_0^i = \frac{ES_1^i}{1 + r + \beta_i(E\mathbf{R}^M - r)}.$$

This formula generalizes the usual deterministic discounting formula, according to which the future payment is discounted by the factor $1/(1 + r)$. In the random case, the appropriate interest rate is $r + \beta_i(E\mathbf{R}^M - r)$ (a "risk-adjusted" interest rate).

8.2 CAPM as a Factor Model

The Structure of Asset Returns Under the CAPM The CAPM implies a special structural property for asset returns, and this property provides a further explanation why beta is the most important measure of risk. To derive this result, consider the random variable

$$\varepsilon_i = R_i - r - \beta_i(\mathbf{R}^M - r). \tag{8.1}$$

We can represent the (random) return on asset i as

$$R_i = r + \beta_i(\mathbf{R}^M - r) + \varepsilon_i. \tag{8.2}$$

By taking the expectation in (8.2) and using the CAPM,

$$ER_i = r + \beta_i(E\mathbf{R}^M - r), \tag{8.3}$$

we find that $E\varepsilon_i = 0$. Further, by taking the covariance of both sides of (8.2) with $\mathbf{R}^M$, we get

$$Cov(R_i, \mathbf{R}^M) = \beta_i Cov(\mathbf{R}^M, \mathbf{R}^M) + Cov(\varepsilon_i, \mathbf{R}^M). \tag{8.4}$$

Using the fact that $Cov(\mathbf{R}^M, \mathbf{R}^M) = Var\mathbf{R}^M$ and the definition of β_i, we find

$$\beta_i Cov(\mathbf{R}^M, \mathbf{R}^M) = Cov(R_i, \mathbf{R}^M).$$

Hence (8.4) implies $Cov(\varepsilon_i, \mathbf{R}^M) = 0$. Thus the term ε_i *has zero expectation and is uncorrelated with* $\mathbf{R}^M$.

Thus we obtain the *one-factor model* for the returns R_i:

$$R_i = ER_i + \beta_i F + \varepsilon_i,\ i = 1, 2, \ldots, N, \tag{8.5}$$

where $F = \mathbf{R}^M - E\mathbf{R}^M$, $EF = E\varepsilon_i = 0$ and $Cov(F, \varepsilon_i) = 0$.

To prove (8.5) we use (8.1) and write (8.5) as

$$R_i = ER_i + \beta_i(\mathbf{R}^M - E\mathbf{R}^M) + R_i - r - \beta_i(\mathbf{R}^M - r).$$

By cancelling out terms, we transform this equation to

$$0 = ER_i + \beta_i(-E\mathbf{R}^M) - r - \beta_i(-r).$$

This is equivalent to (8.3), and so (8.5) is proved.

Remark The random variable $F = \mathbf{R}^M - E\mathbf{R}^M$ (the *factor*) is interpreted here as a "basic source of randomness," which is common for all the asset returns. The residual terms ε_i characterize "specific sources of randomness" (specific for each particular asset $i = 1, 2, \ldots, N$).

Systematic and Specific Risk We have shown that

$$R_i = ER_i + \beta_i(\mathbf{R}^M - E\mathbf{R}^M) + \varepsilon_i,$$

where

$$E\varepsilon_i = Cov(\mathbf{R}^M, \varepsilon_i) = 0.$$

This yields

$$VarR_i = \beta_i^2 Var\mathbf{R}^M + Var\,\varepsilon_i$$

because the variance of the sum of uncorrelated random variables is equal to the sum of their variances. In this formula, the term $\beta_i^2 Var\mathbf{R}^M$ is called *systematic risk*. It reflects uncertainty associated with the market as a whole. This risk cannot be reduced by portfolio diversification, since every asset with non-zero beta contains this risk. The second part, $Var\,\varepsilon_i$ is called *non-systematic* (or *specific*, or *idiosyncratic*) risk. The latter kind of risk is uncorrelated with the market and under certain conditions (see below), it can be reduced by portfolio diversification: by including in the portfolio a large number of different assets.

Portfolio Diversification Assume that there are sufficiently many assets $i = 1, 2, \ldots, N$ in the market, the residual terms ε_i for these assets are uncorrelated with each other, and the variances $Var\,\varepsilon_i$ are bounded by some constant C. Recall that

$$R_i = ER_i + \beta_i(\mathbf{R}^M - E\mathbf{R}^M) + \varepsilon_i,$$

where $E\varepsilon_i = Cov(\mathbf{R}^M, \varepsilon_i) = 0$. Therefore for each normalized portfolio $x = (x_1, \ldots, x_N)$ of N risky assets, we have

$$R_x = ER_x + \beta_x(\mathbf{R}^M - E\mathbf{R}^M) + \varepsilon_x, \tag{8.6}$$

where

$$\beta_x = \sum_i \beta_i x_i = \sum_i x_i \frac{Cov(R_i, \mathbf{R}^M)}{Var\mathbf{R}^M} = \frac{Cov(R_x, \mathbf{R}^M)}{Var\mathbf{R}^M}$$

and $\varepsilon_x = \sum_i \varepsilon_i x_i$. Clearly $E\varepsilon_x = Cov(\mathbf{R}^M, \varepsilon_x) = 0$, and so it is natural to regard the variance of the residual term ε_x in formula (8.6) as *the specific risk of the portfolio* x.

If we construct a portfolio, for example, of the form $x = (\frac{1}{N}, \ldots, \frac{1}{N})$, then the specific risk for this portfolio will be equal to

$$Var\,\varepsilon_x = Var\sum_{i=1}^{N} \frac{1}{N}\varepsilon_i = \sum_{i=1}^{N} Var(\frac{1}{N}\varepsilon_i) = \sum_{i=1}^{N} \frac{1}{N^2} Var(\varepsilon_i) \leq \frac{1}{N^2} CN = \frac{C}{N}$$

because the residuals ε_i are uncorrelated. This shows that $Var\,\varepsilon_x$ becomes small when N is large ($\frac{C}{N} \to 0$ as $N \to \infty$).

8.3 Applying Theory to Practice: Sharpe's and Jensen's Tests

Applying Theory to Practice: Evaluating the Performance of a Portfolio The above theory can be applied in practice for the evaluation of a portfolio performance. Let us outline the general approach which is used for this purpose. Consider a financial market with N risky assets ($i = 1, 2, \ldots, N$) and a risk-free asset—cash ($i = 0$). Suppose we have to evaluate the performance of some given portfolio $\mathbf{x}$ based on statistical data regarding this portfolio and the market as a whole. These data are available for a certain period of time, say n years. Assume that we have:

- The n year record of rates of return on the portfolio $\mathbf{x}$:

$$R(1), R(2), \ldots, R(n).$$

- The n year record of rates of return on the market portfolio $\mathbf{x}^M$:

$$R^M(1), R^M(2), \ldots, R^M(n).$$

- The n year record of interest rates:

$$r(1), r(2), \ldots, r(n).$$

Empirical Expectations, Variances and Covariances Based on these data, we proceed as follows. We assume that cash is a risk-free asset whose rate of return is equal to the average of the interest rates $r(t)$, $t = 1, \ldots, n$:

$$r = \frac{1}{n}[r(1) + r(2) + \ldots + r(n)].$$

We compute the averages

$$ER = \frac{1}{n}[R(1) + R(2) + \ldots + R(n)],$$
$$ER^M = \frac{1}{n}[R^M(1) + R^M(2) + \ldots + R^M(n)]$$

and use them as proxies for the expected values of the returns on the portfolios $\mathbf{x}$ and $\mathbf{x}^M$, respectively. Also, we estimate the variances $Var\ R$, $Var\ R^M$ and the covariance $Cov(R, R^M)$ of these returns by using the following formulas.

The formula for estimating the variance of a random variable R based on the observed values $R(1),\ldots,R(n)$ of R:

$$VarR = \frac{1}{n-1}\sum_{t=1}^{n}(R(t) - ER)^2,$$

where $ER = \frac{1}{n}[R(1) + \ldots + R(n)]$.

The formula for estimating the covariance between two random variables R_1 and R_2 based on the observed values $R_1(1), \ldots, R_1(n)$ and $R_2(1), \ldots, R_2(n)$:

$$Cov(R_1, R_2) = \frac{1}{n-1} \sum_{t=1}^{n} (R_1(t) - ER_1)(R_2(t) - ER_2),$$

where ER_1 and ER_2 are the estimates for the expectations

$$ER_1 = \frac{1}{n} \sum_{t=1}^{n} R_1(t), \; ER_2 = \frac{1}{n} \sum_{t=1}^{n} R_2(t).$$

Having computed $Var\, R^M$ and $Cov(R, R^M)$ by using the above formulas, we calculate the beta of the portfolio:

$$\beta = \frac{Cov(R, R^M)}{Var(R^M)}.$$

Jensen's Test We describe two tests for the performance of a portfolio. In the first test, we evaluate *the Jensen index*: the number J satisfying

$$ER - r = J + \beta(ER^M - r).$$

Clearly, J is the difference between the left-hand side and the right-hand side of (the empirical version of) the CAPM equation. Here ER, ER^M and β are the statistical estimates of the corresponding characteristics we obtained based on observations. The theoretical value of J is zero, as long as the CAPM holds.

If $J > 0$ *(resp.* $J < 0$*) then we conclude that the portfolio* $\mathbf{x}$ *yields better (resp. worse) expected returns than those predicted by the CAPM.*

Sharpe's Test As we know, the *Sharpe ratio* for a portfolio $\mathbf{x} = (x_0, \ldots, x_N)$ is given by

$$\frac{m_{\mathbf{x}} - r}{\sigma_{\mathbf{x}}},$$

where $m_{\mathbf{x}}$ and $\sigma_{\mathbf{x}}$ are the expectation and the standard deviation of the return on the portfolio $\mathbf{x}$, and r is the return on the risk-free asset. Replacing $m_{\mathbf{x}}$ and $\sigma_{\mathbf{x}}$ by their empirical values, ER and $\sqrt{VarR}$ (obtained above based on the statistical data), we compute the estimate for the Sharpe ratio of the portfolio $\mathbf{x}$ as:

$$\rho = \frac{ER - r}{\sqrt{VarR}}.$$

Analogously, we compute the estimate for the Sharpe ratio of the market portfolio $\mathbf{x}^M$:

$$\rho^M = \frac{ER^M - r}{\sqrt{VarR^M}}.$$

According to the theory of the CAPM, the market portfolio $\mathbf{x}^M$ is efficient, and all efficient portfolios have the same Sharpe ratio.

The Sharpe test is performed as follows. We fix (in advance) some percentage level γ, say $\gamma = 5\,\%$ and compare the Sharpe ratios ρ^M and ρ in order to check whether the former is greater than the latter by more than 5 %, i.e.

$$\frac{\rho^M - \rho}{\rho^M} > 0.05.$$

If this inequality holds, then we conclude that the performance of the portfolio $\mathbf{x}$ is not satisfactory: its efficiency measured in terms of the Sharpe ratio is not large enough. If the opposite inequality holds, it can be concluded that the portfolio $\mathbf{x}$ does have a sufficiently high level of efficiency.

Numerical examples of application of the above tests will be considered in Chap. 10.

9 Factor Models and the Ross-Huberman APT

9.1 Single- and Multi-Factor Models

CAPM as a Single-Factor Model It was shown in the previous chapter that the CAPM equation implies that the return R_i on every asset $i = 1, \ldots, N$ can be represented in the form

$$R_i = ER_i + \beta_i F + \varepsilon_i \tag{9.1}$$

where

$$EF = E\varepsilon_i = 0 \text{ and } Cov(F, \varepsilon_i) = 0. \tag{9.2}$$

Definition If there exist random variables F and ε_i $(i = 1, \ldots, N)$ and non-random numbers β_i $(i = 1, 2, \ldots, N)$ such that (9.1) and (9.2) hold, then we say that the financial market under consideration can be described by a *single-factor model* for asset returns. The random variable F is called the (risk) *factor* and is interpreted as a "basic source of randomness," which is common for all the asset returns. It is normally assumed that $VarF > 0$. The *residual terms* (*residuals*) ε_i characterize "specific sources of randomness" (specific for each asset i).

Based on the CAPM relation, we derived (9.1) and (9.2) for the factor F that was defined as

$$F = \mathbf{R}^M - E\mathbf{R}^M.$$

I.V. Evstigneev et al., *Mathematical Financial Economics*, Springer Texts in Business and Economics, DOI 10.1007/978-3-319-16571-4_9

Here, $\mathbf{R}^M$ stands for the return on the market portfolio $\mathbf{x}^M$. Recall that, by the definition of the market portfolio $\mathbf{x}^M$, every coordinate x_i^M of the vector

$$\mathbf{x}^M = (x_0^M, x_1^M, \ldots, x_N^M)$$

represents the fraction of total market wealth invested (by all the market participants) in asset i.

Multifactor Models A single-factor model is a very crude approximation to reality, and this is one of the reasons why the conventional CAPM theory leads to a number of paradoxes and anomalies. A more realistic approach is based on multifactor models.

Definition We say that the financial market under consideration can be described by *a K-factor model* if there exist K random variables $F_1, F_2, \ldots, F_K$ (*factors*) with $VarF_k > 0$ and N random variables $\varepsilon_1, \varepsilon_2, \ldots, \varepsilon_N$ (*residuals*) such that the following conditions hold:

(a) for each $i = 1, 2, \ldots, N$, the random variable R_i—the return on asset i—can be represented as

$$R_i = ER_i + c_{i1}F_1 + \ldots + c_{iK}F_K + \varepsilon_i,$$

where $c_{i1}, \ldots, c_{iK}$ are non-random numbers (c_{ik} is the *exposure* or *sensitivity* of asset i to factor k);

(b) the factors $F_1, \ldots, F_K$ are mutually uncorrelated:

$$Cov(F_k, F_l) = 0 \text{ for all } k \neq l;$$

(c) the residuals are mutually uncorrelated:

$$Cov(\varepsilon_i, \varepsilon_j) = 0 \text{ for all } i \neq j,$$

and uncorrelated with each factor:

$$Cov(\varepsilon_i, F_k) = 0 \text{ for all } i \text{ and } k;$$

(d) the expected values of the factors and of the residuals are equal to zero:

$$EF_k = E\varepsilon_i = 0 \text{ for all } k \text{ and } i.$$

[Note that $k, l = 1, 2, \ldots, K$ while $i, j = 1, 2, \ldots, N$.]

Remark When analyzing factor models, we will deal with N assets, but we will not assume—in contrast with the previous study of the Markowitz model with N assets—that all of the assets are necessarily risky (the covariance matrix $V = (Cov(R_i, R_j))$ will not necessarily be positive definite). Therefore the analysis will comprise, in particular, the Markowitz model with a risk-free asset considered in the previous chapters.

K-Factor Model: Vector Notation By using the vector notation

$$R = (R_1, \ldots, R_N),$$

$$m = (m_1, \ldots, m_N) = (ER_1, \ldots, ER_N),$$

$$c_k = (c_{1k}, \ldots, c_{Nk}),\ \varepsilon = (\varepsilon_1, \ldots, \varepsilon_N),$$

we can write the relation

$$R_i = ER_i + c_{i1}F_1 + \ldots + c_{iK}F_K + \varepsilon_i \text{ for all } i,$$

involved in the definition of a K-factor model, as follows:

$$R = m + c_1F_1 + \ldots + c_KF_K + \varepsilon.$$

Recall that $EF_k = E\varepsilon_i = 0$ for all k, i and

$$Cov(F_k, F_l) = Cov(\varepsilon_i, \varepsilon_j) = Cov(\varepsilon_i, F_k) = 0$$

for all $k \neq l$, $i \neq j$ and all i, k.

9.2 Exact Factor Pricing

Exact Factor Model Let us begin the analysis of the K-factor model by imposing a very strong simplifying assumption. Let us assume for the moment that $\varepsilon_i = 0$ for all i. This means that there are no residual terms, and the returns on all the assets $i = 1, 2, \ldots, N$ *can exactly be expressed through the factors* $F_1, \ldots, F_K$. By using the vector notation, we can write:

$$R = m + c_1F_1 + \ldots + c_KF_K.$$

Return on a Portfolio in the Exact Factor Model We know that the return R_x on a portfolio $x = (x_1, \ldots, x_N)$ can be computed by the formula

$$R_x = \langle R, x\rangle = \sum_{i=1}^{N} R_i x_i.$$

If the exact factor equation

$$R = m + c_1 F_1 + \ldots + c_K F_K$$

holds, we can compute R_x as follows:

$$R_x = \langle R, x \rangle = \langle m, x \rangle + \sum_{k=1}^{K} \langle c_k, x \rangle F_k .$$

The coefficient $\langle c_k, x \rangle$ is the *exposure of the portfolio x to factor k*. We have

$$\langle c_k, x \rangle = c_{1k} x_1 + \ldots + c_{Nk} x_N ,$$

and so $\langle c_k, x \rangle$ is the sum

$$\sum_{i=1}^{N} (\text{exposure of asset } i \text{ to factor } k) \times x_i .$$

Arbitrage The following definition is very important. Various versions of it play key roles in many models of mathematical finance. This concept will appear in this book many times.

Definition We say that there is an *arbitrage opportunity* in the financial market if there exists a self-financing portfolio x such that

$$R_x \geq 0 \text{ with probability } 1$$

and

$$R_x > 0 \text{ with strictly positive probability.}$$

Recall that a portfolio $x = (x_0, \ldots, x_N)$ is called self-financing if $\langle e, x \rangle = x_1 + \ldots + x_N = 0$.

Arbitrage: Interpretation To understand what the existence of an arbitrage opportunity means, let us recall [see (1.4)] that for a self-financing portfolio x the return R_x is equal to the value of the portfolio at the terminal date 1,

$$R_x = w_1 ,$$

because the initial value $w_0 = x_1 + \ldots + x_N$ is zero. According to the definition of an arbitrage opportunity, $w_1 \geq 0$ with probability one and $w_1 > 0$ with strictly positive probability. Consequently, starting with zero initial wealth, one can construct a portfolio which has always non-negative value and, in some situations, strictly positive value. Roughly speaking, an arbitrage opportunity is a way of getting something from nothing!

No Arbitrage Hypothesis It is natural to impose the following assumption.

(NA) The asset market under consideration does not allow for arbitrage opportunities.

In reality, if arbitrage opportunities appear, the market immediately reacts to this by a change in asset prices eliminating arbitrage.

The No Arbitrage Hypothesis in the Exact Factor Model Suppose the market is described by the exact factor model:

$$R = m + c_1 F_1 + \ldots + c_K F_K.$$

Let us say that $x = (x_1, \ldots, x_N)$ is a *portfolio without factor exposure* if

$$\langle x, c_1 \rangle = \langle x, c_2 \rangle = \ldots = \langle x, c_K \rangle = 0.$$

Clearly, for such a portfolio, $R_x = \langle x, m \rangle$, and so the return on x is not random.

Proposition 9.1 *Under (NA), every self-financing portfolio without factor exposure yields zero return.*

Proof The return on a portfolio x without factor exposure is non-random and is equal to $\langle x, m \rangle$. Suppose $\langle x, m \rangle \neq 0$ and x is self-financing, i.e., $\langle e, x \rangle = 0$. We may assume without loss of generality that $\langle x, m \rangle > 0$. If $\langle x, m \rangle < 0$, we can replace x by $-x$, which is also a self-financing portfolio without factor exposure, and for which $\langle -x, m \rangle > 0$. Thus x is a self-financing portfolio with strictly positive return. This contradicts assumption (**NA**).

The proof is complete. □

The Exact Factor Pricing Theorem A central result related to the exact factor model is the following theorem.

Theorem 9.1 (The Exact Factor Pricing Theorem) *There exist numbers* $\lambda_0, \lambda_1, \ldots, \lambda_K$ *such that*

$$ER_i = \lambda_0 + \lambda_1 c_{i1} + \ldots + \lambda_K c_{iK}, \text{ for each } i = 1, 2, \ldots, N.$$

By using vector notation, we can write this formula as

$$ER = \lambda_0 e + \lambda_1 c_1 + \ldots + \lambda_K c_K,$$

where $c_k = (c_{1k}, c_{2k}, \ldots, c_{Nk})$ and $e = (1, \ldots, 1)$. The numbers $\lambda_1, \ldots, \lambda_K$ are called *factor risk premia*. Theorem 9.1 says that the expected return ER_i on each asset i can be expressed as a linear function $\lambda_0 + \lambda_1 c_{i1} + \ldots + \lambda_K c_{iK}$ of factor exposures $c_{i1}, c_{i2}, \ldots, c_{iK}$ of this asset, with coefficients $\lambda_0, \lambda_1, \ldots, \lambda_K$ *independent* of i.

Proof of Theorem 9.1 By Proposition 9.1, every self-financing portfolio x without factor exposure has zero expected return. This means that every vector x which is a solution to the system of linear equations

$$\langle e, x\rangle = 0, \langle c_1, x\rangle = 0, \ldots, \langle c_K, x\rangle = 0$$

is a solution to the linear equation $\langle m, x\rangle = 0$. By a theorem in linear algebra (see Appendix A, Proposition 5), this implies that the vector $m = ER$ is a linear combination $\lambda_0 e + \lambda_1 c_1 + \ldots + \lambda_K c_K$ of $e, c_1, \ldots, c_K$, which proves the theorem.

Remark As in the CAPM, in Theorem 9.1 we do not consider the asset prices S_0^i and S_1^i explicitly—only the returns R_i. However, the result can be reformulated in terms of the prices because $R_i = (S_1^i - S_0^i)/S_0^i$, and so $ER_i = (ES_1^i - S_0^i)/S_0^i$.

What If One of the Assets $i = 1, 2, \ldots, N$ Is Risk-Free? Consider an exact factor model

$$R = m + c_1 F_1 + \ldots + c_K F_K. \tag{9.3}$$

If the no arbitrage hypothesis holds, we have the exact factor pricing formula

$$m = \lambda_0 e + c_1 \lambda_1 + \ldots + c_K \lambda_K. \tag{9.4}$$

Proposition 9.2 *If one of the assets $i = 1, 2, \ldots, N$, say $i = 1$, is risk-free with non-random return r, then $\lambda_0 = r$.*

Proof Equation (9.3) can be written for each coordinate $i = 1, 2, \ldots, N$ of the vectors involved; let us write it for $i = 1$:

$$r = r + c_{11} F_1 + \ldots + c_{1K} F_K.$$

This implies

$$0 = c_{11} F_1 + \ldots + c_{1K} F_K.$$

Denoting the expression $c_{11}F_1 + \ldots + c_{1K}F_K$ by G, and multiplying G by any F_j, $j = 1, 2, \ldots, K$, we find

$$0 = E(GF_j) = c_{1j}EF_j^2 = c_{1j}\,VarF_j$$

because $EF_i = EF_iF_j = Cov(F_i, F_j) = 0$ for all $i \neq j$. Thus $c_{1j} = 0$ since $VarF_j > 0$.

Writing relation (9.4) for the first coordinate of the vectors involved, we obtain

$$r = \lambda_0 + c_{11}\lambda_1 + \ldots + c_{1K}\lambda_K = \lambda_0$$

because, as we have proved, $c_{1j} = 0$ for all j. This proves the proposition. □

Is CAPM an Exact Factor Model? We would like to clarify relationships between the CAPM (introduced as single-factor model) and the exact factor models.

First of all, is CAPM an exact factor model? Clearly, the answer is "No". Indeed, the CAPM theory leads to the relation

$$R_i = ER_i + \beta_i F + \varepsilon_i,$$

where the residual ε_i is, generally, non-zero.

Does CAPM Imply Exact Factor Pricing? Yes, it does. Indeed, we have shown that the CAPM equation implies

$$R_i = ER_i + \beta_i(\mathbf{R}^M - E\mathbf{R}^M) + \varepsilon_i,$$

where the random variable $F = \mathbf{R}^M - E\mathbf{R}^M$ plays the role of a factor, so that $Cov(F, \varepsilon_i) = EF = E\varepsilon_i = 0$. Here, the coefficient

$$\beta_i = Cov(R_i, \mathbf{R}^M)/Var\mathbf{R}^M,$$

the beta of the asset i, plays the role of the exposure of asset i to the single factor F. But the CAPM equation says that

$$ER_i = r + \beta_i(E\mathbf{R}^M - r),$$

where r is the return on the risk-free asset. By setting

$$\lambda_0 = r \text{ and } \lambda_1 = E\mathbf{R}^M - r,$$

we obtain the formula of exact factor pricing:

$$ER_i = \lambda_0 + \beta_i\lambda_1.$$

9.3 Ross-Huberman APT: Model Description

APT "APT" stands for "Arbitrage Pricing Theory." It was proposed by Ross (1976) and Huberman (1982) as an alternative to the CAPM, and then developed by many others. The idea of APT is as follows. Consider a general, not necessarily exact, K-factor model

$$R = m + c_1 F_1 + \ldots + c_K F_K + \varepsilon,$$

where the vector of residuals $\varepsilon = (\varepsilon_1, \ldots, \varepsilon_N)$ is not necessarily zero. In this model, one cannot guarantee that the exact factor pricing formula

$$m = \lambda_0 e + \lambda_1 c_1 + \ldots + \lambda_K c_K,$$

holds. (Recall that m is the vector of *expected returns*, and the coordinates c_{ik} of the vectors c_k are the *exposures* of assets i to factors k.) However, according to the APT, this formula still holds **approximately**,

$$m \approx \lambda_0 e + \lambda_1 c_1 + \ldots + \lambda_K c_K,$$

and it becomes precise in the limit as $N \to \infty$, i.e., when the number of assets in the market tends to infinity.

"Large" Asset Market Formally, APT deals with a sequence of markets M^N, $N = 1, 2, \ldots$ such that, in the market M^N, N assets $i = 1, 2, \ldots, N$ with random returns R_i are traded. The random variables R_i are defined for each $i = 1, 2, \ldots$. When $N \to \infty$, this provides a model for a "large" asset market.

We will assume that each of the markets M^N can be described by a K-factor model with the same factors $F_1, F_2, \ldots, F_K$, so that

$$R_i = ER_i + c_{i1} F_1 + \ldots + c_{iK} F_K + \varepsilon_i \text{ for all } i = 1, \ldots, N,$$

where $Cov(F_k, F_l) = Cov(\varepsilon_i, \varepsilon_j) = Cov(\varepsilon_i, F_k) = 0$ and $EF_k = E\varepsilon_i = 0$. Additionally, we suppose that the variances $Var\varepsilon_i$ (for all $i = 1, 2, \ldots$) are bounded by some constant C.

Asymptotic Arbitrage We say that there is an *asymptotic arbitrage opportunity* if there exists a sequence of natural numbers $N_1, N_2, \ldots$ and a sequence of self-financing portfolios $x(N_1), x(N_2), \ldots$ in the markets $M^{N_1}, M^{N_2}, \ldots$ such that

$$ER_{x(N_j)} \to \infty \text{ and } VarR_{x(N_j)} \to 0$$

as $j \to \infty$. (One can become infinitely rich with zero risk starting from zero wealth.)

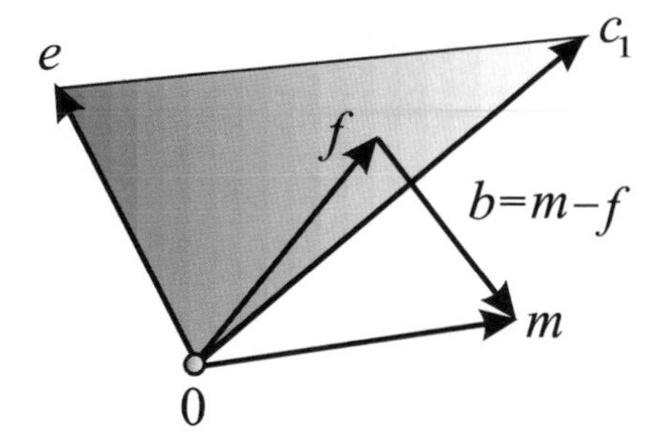

Fig. 9.1 The vector of factor pricing errors

We will introduce a hypothesis that rules out asymptotic arbitrage, similar to hypothesis (**NA**) considered before.

(**NAA**) There are no asymptotic arbitrage opportunities.

Factor Pricing Errors To formulate the main result of APT, let us denote by $f = f(N)$ that vector in the N-dimensional space R^N which can be represented as

$$f = \lambda_0 e + \lambda_1 c_1 + \ldots + \lambda_K c_K$$

for some $\lambda_0, \ldots, \lambda_K$ and for which the *distance*[1] between f and m is a minimum. Recall that $m = (ER_1, \ldots, ER_N)$ is the vector of expected returns on assets $i = 1, 2, \ldots, N$. The vector $b = b(N)$, where

$$b = m - f,$$

is called *the vector of factor pricing errors for assets* $i = 1, 2, \ldots, N$ in the market M^N. The norm

$$\rho_N = \|b(N)\|$$

is called the *factor pricing error for the market* M^N. A geometric illustration is given in Fig. 9.1.

[1]The *distance* between two vectors $a = (a_1, \ldots, a_N)$ and $b = (b_1, \ldots, b_N)$ is equal to $\|a - b\|$, where $\|\cdot\|$ stands for the norm of a vector. The *norm* (the length) of a vector $c = (c_1, \ldots, c_N)$ is defined by $\|c\| = \sqrt{c_1^2 + \ldots + c_N^2}$. For these and other mathematical notions involved in this chapter see Appendix A.

9.4 Formulation and Proof of the Main Result

The Ross-Huberman Theorem: Formulation Consider the model of a "large" asset market described in terms of an increasing sequence of markets M^N satisfying the assumptions introduced in the previous sections. A central result of APT is the following theorem.

Theorem 9.3 *The factor pricing errors ρ_N in the sequence of markets M^N are bounded, i.e., there exists a constant H such that*

$$\rho_N \leq H \text{ for all } N.$$

The Ross-Huberman Theorem: Discussion The meaning of this result is as follows. We can write

$$\rho_N = \|b(N)\|,$$

where

$$b(N) = (b_1(N), \ldots, b_N(N))$$

is the vector of factor pricing errors for assets $i = 1, 2, \ldots, N$ in the market M^N. The inequality $\rho_N \leq H$ implies

$$\sum_{i=1}^{N} b_i(N)^2 \leq H^2.$$

This means that *if N is large, then for most of the assets $i = 1, 2, \ldots, N$, the factor pricing errors $b_i(N)$ are small.* Indeed, if N is large, then there are many summands in the sum $\sum_{i=1}^{N} b_i(N)^2$, but the sum is bounded by a constant which does not depend on N. In particular,

$$\frac{1}{N}\sum_{i=1}^{N} b_i(N)^2 \to 0 \text{ as } N \to \infty,$$

i.e., the average squared error $b_i(N)$ tends to zero.

Proof of Theorem 9.3

1st step. Recall that, for each of the markets M^N, we have

$$R = m + \sum_{k=1}^{K} c_k F_k + \varepsilon.$$

Let f be the vector of the form

$$f = \lambda_0 e + \lambda_1 c_1 + \ldots + \lambda_K c_K$$

for which the distance between f and m is a minimum. Put $c_0 = e$. The vector f is the *projection* of m on the *linear span* of the vectors $c_0, c_1, \ldots, c_K$. The vector $b = m - f$ has the following property:

$$\langle b, c_k \rangle = \langle m - f, c_k \rangle = 0, \; k = 0, \ldots, K,$$

that is, the vector $b = m - f$ is *orthogonal*[2] to each of the vectors c_k, $k = 0, 1, \ldots, K$. The equalities

$$\langle m - f, c_k \rangle = 0, k = 0, \ldots, K,$$

imply

$$\langle b, f \rangle = 0 \tag{9.5}$$

because f is a linear combination of $c_0, c_1, \ldots, c_K$ (see the above Fig. 9.1).

2nd step. Consider the vector $b = (b_1, \ldots, b_N) = m - f$ of factor pricing errors. We can regard b (as any other vector) as a portfolio of assets: b_i is the amount invested in asset i. We list some properties of this portfolio which are of importance for the proof of our theorem.

(i) The *portfolio b is self-financing* because

$$\sum_{i=1}^{N} b_i = \langle b, e \rangle = \langle b, c_0 \rangle = 0.$$

(ii) By virtue of the K-factor relation

$$R = m + \sum_{k=1}^{K} c_k F_k + \varepsilon,$$

the return on the portfolio b is equal to

$$\langle b, R \rangle = \langle b, m \rangle + \langle b, \varepsilon \rangle, \tag{9.6}$$

because $\langle b, c_k \rangle = 0$ for each $k = 1, 2, \ldots, K$.

[2] Two vectors a and b are called *orthogonal* if $\langle a, b \rangle = 0$. Geometrically, this means that the angle between a and b is the right angle 90°.

(iii) By virtue of (9.6), the expected return on the portfolio b is equal to $E\langle b, R\rangle = \langle b, m\rangle$, since $E\varepsilon = 0$. Further,

$$\langle b, m\rangle = \langle b, m\rangle - \langle b, f\rangle = \langle b, m - f\rangle = \langle b, b\rangle = \|b\|^2,$$

where the first equality holds because $\langle b, f\rangle = 0$ [see (9.5)]. Thus the *expected return on the portfolio b is*

$$E\langle b, R\rangle = \langle b, m\rangle = \|b\|^2.$$

(iv) Also, it follows from (9.6) that the *variance of the return on the portfolio b satisfies*

$$Var\langle b, R\rangle = Var\langle b, \varepsilon\rangle = \sum_{i=1}^{N} b_i^2 Var\, \varepsilon_i \leq C\,\|b\|^2,$$

because $\varepsilon_1, \ldots, \varepsilon_N$ are uncorrelated and $Var\,\varepsilon_i \leq C$.

3rd step. We have shown that the portfolio $b = m - f$ is self-financing, its expected return ER_b is $\|b\|^2$ and the variance $VarR_b$ of its return does not exceed $C\,\|b\|^2$. Consider any number $\gamma > 0$ and define a portfolio x by $x = \gamma b$. Then x is self-financing, and we have

$$ER_x = ER_{\gamma b} = \gamma ER_b = \gamma\,\|b\|^2,$$

$$VarR_x = VarR_{\gamma b} = \gamma^2 VarR_b \leq C\gamma^2\|b\|^2.$$

Define the number γ as follows

$$\gamma = \|b\|^{-\frac{3}{2}}.$$

Then

$$ER_x = \|b\|^{\frac{1}{2}} \tag{9.7}$$

and

$$VarR_x \leq C\,\|b\|^{-1}. \tag{9.8}$$

Step 4. Now let $b = b(N)$ be the vector of factor pricing errors for assets $i = 1, \ldots, N$ in the market M^N and $x = x(N)$ the self-financing portfolio of assets $i = 1, 2, \ldots, N$ for which relations (9.7) and (9.8) hold. Suppose the sequence

$||b(N)||$ is not bounded. This means that $\|b(N)\| \to \infty$ for some sequence $N = N_1, N_2, \ldots$. Then we can see from (9.7) and (9.8) that $ER_{x(N)} \to \infty$, $Var(R_{x(N)}) \to 0$, for $N = N_1, N_2, \ldots$, which yields an asymptotic arbitrage opportunity. This is a contradiction.

Consequently, the sequence $\rho_N = \|b(N)\|$ is bounded. This completes the proof.

Problems and Exercises I

10

Question 10.1 (Tangency Portfolio and Sharpe Ratio)

(a) Show that the Sharpe ratio $\rho_{\mathbf{v}}$ for the tangency portfolio $\mathbf{v} = (0, v)$ can be computed by the formula

$$\rho_{\mathbf{v}} = \rho_v = \sqrt{H},$$

where $H = Cr^2 - 2Ar + B$. Recall that throughout the book we use the following notation:

$$A = \langle e, Wm \rangle, \ B = \langle m, Wm \rangle, \ C = \langle e, We \rangle, \ D = BC - A^2, \ W = V^{-1}.$$

(b) Prove that the Sharpe ratio

$$\rho_x = \frac{m_x - r}{\sigma_x}$$

attains its unique maximum on the set of all portfolios $x = (x_1, \ldots, x_N) \in R^N$, containing only risky assets, at the portfolio $v \in R^N$, where v is the risky component of the tangency portfolio: $\mathbf{v} = (0, v)$. To obtain the result, proceed as follows:

(i) observe that it is sufficient to look for a maximum of ρ_x only among efficient portfolios x_τ^*, $\tau \geq 0$;

(ii) use the formulas for $\mu = \mu(\tau) = m_{x_\tau^*}$ and $\sigma^2 = \sigma^2(\tau) = \sigma^2_{x_\tau^*}$:

$$\mu(\tau) = \frac{A}{C} + \frac{\tau}{2}\frac{D}{C}, \text{ and } \sigma^2(\tau) = \frac{1}{C} + \frac{\tau^2}{4}\frac{D}{C}; \tag{10.1}$$

I.V. Evstigneev et al., *Mathematical Financial Economics*, Springer Texts in Business and Economics, DOI 10.1007/978-3-319-16571-4_10

(iii) consider the ratio

$$\rho_{x_\tau^*} = \frac{\mu(\tau) - r}{\sigma(\tau)}$$

as a function of $\tau \geq 0$ and show that this function is strictly less than $\sqrt{H}$ for all $\tau \neq \bar{\tau}$ and is equal to $\sqrt{H}$ for $\tau = \bar{\tau}$.

(c) Draw a diagram depicting the efficient frontier (in the $\sigma_x - m_x$ plane) for the market containing N risky assets $i = 1, 2, .., N$. In this diagram, indicate what is the Sharpe ratio of an efficient portfolio x_τ^*. Explain why it follows from the result obtained in (b) that the straight line—the efficient frontier of the market including the risk-free asset—is tangent to the hyperbola depicting the efficient frontier of the market of N risky assets, the point of tangency being (σ_v, m_v).

Answer

(a) In Chap. 6, the following formulas were obtained for the variance σ^2 and the expectation μ of the return on the efficient portfolio $\mathbf{x}_\tau^*$:

$$\sigma^2 = \frac{\tau^2}{4} H, \text{ and } \mu = r + \frac{\tau}{2} H. \tag{10.2}$$

Computing the Sharpe ratio, we find:

$$\frac{\mu - r}{\sigma} = \frac{r + \frac{\tau}{2} H - r}{\frac{\tau}{2}\sqrt{H}} = \sqrt{H}.$$

Thus *the Sharpe ratio is the same and is equal to* $\sqrt{H}$ for all efficient portfolios $\mathbf{x}_\tau^*$ (in the market with a risk-free asset). In particular, the above formula is applicable to the tangency portfolio $\mathbf{v}$ because $\mathbf{v} = \mathbf{x}_{\bar{\tau}}^*$, where $\bar{\tau} = (A - rC)^{-1}$.

(b) We have $\mathbf{v} = (0, v)$, and so, for the portfolio v,

$$\rho_v = \sqrt{H}$$

because $\sigma_{\mathbf{v}} = \sigma_v$ and $m_{\mathbf{v}} = m_v$. Thus it suffices to show that

$$\frac{\mu - r}{\sigma} < \sqrt{H}$$

for $\mu = m_x$ and $\sigma = \sigma_x$ corresponding to any portfolio of risky assets $x \neq v$.

(i) It is sufficient to prove the inequality $\dfrac{m_x - r}{\sigma_x} < \sqrt{H}$ only for *efficient* portfolios $x \neq v$. Indeed, if a portfolio y is not efficient, then there is an efficient portfolio x for which the variance σ_x is less than σ_y and $m_x \geq m_y$

(we can define x as the normalized portfolio which satisfies $m_x \geq m_y$ and has the smallest variance). Then

$$\frac{m_y - r}{\sigma_y} \leq \frac{m_x - r}{\sigma_x} < \sqrt{H},$$

as long as we know that the last inequality is valid for all efficient x.

(ii) Since for an efficient portfolio $x = x_\tau^*$, $\mu(\tau) = m_{x_\tau^*}$ and $\sigma(\tau) = \sigma_{x_\tau^*}$ (with $\mu(\tau)$ and $\sigma(\tau)$ defined in the question), it is sufficient to prove that

$$\frac{\mu(\tau) - r}{\sigma(\tau)} < \sqrt{H}$$

for each $\tau \neq \bar{\tau}$, or, equivalently,

$$(\mu(\tau) - r)^2 - H\sigma(\tau)^2 < 0, \ \tau \neq \bar{\tau}.$$

(iii) Denote the left-hand side of this inequality by $\phi = \phi(\tau)$. We know that, for $\tau = \bar{\tau}$, the value of the function ϕ is zero because $\bar{\mu} = \mu(\bar{\tau})$ and $\bar{\sigma} = \sigma(\bar{\tau})$ are the values of the variance and the expected return corresponding to the tangency portfolio. Specifically,

$$\mu(\bar{\tau}) = r + \frac{\bar{\tau}}{2}H, \ \sigma^2(\bar{\tau}) = \frac{\bar{\tau}^2}{4}H \tag{10.3}$$

because $x_{\bar{\tau}}^* = v$.

Let us compute the derivative $\phi' = \phi'(\tau)$ of the function

$$\phi = (\mu - r)^2 - H\sigma^2.$$

We write

$$\phi' = 2(\mu - r)\mu' - H(\sigma^2)' =$$

$$(\mu - r)\frac{D}{C} - H\tau\frac{D}{2C} \tag{10.4}$$

by virtue of (10.1). For $\tau = \bar{\tau}$, this derivative is zero, since $\bar{\mu} = r + \bar{\tau}H/2$ (see (10.3)).

Let us now compute the second derivative of ϕ. From (10.4), we can see that

$$\phi'' = \frac{D}{C}[\mu' - \frac{H}{2}],$$

where

$$\frac{\mu'}{2} - \frac{H}{2} = \frac{1}{2}[\frac{D}{C} - H]$$

in view of (10.1). The number $\dfrac{D}{C} - H$ is negative because

$$HC - D = C(Cr^2 - 2Ar + B) - (BC - A^2) = C^2r^2 - 2ArC + A^2$$
$$= (Cr - A)^2 > 0.$$

Thus $\phi(\tau)$ is a strictly concave function, its derivative is zero at $\tau = \bar{\tau}$, and hence $\phi(\tau)$ attains its unique maximum at $\tau = \bar{\tau}$, which completes the proof.

(c) In Fig. 10.1, the curve (hyperbola) is the efficient frontier in the $\sigma_x - m_x$ plane for the market containing N risky assets $i = 1, 2, \ldots, N$.

The straight line emanating from the point $(0, r)$ is the efficient frontier for the market containing a risk-free asset. For all points (σ, μ) in this line, the ratio $(\mu - r)/\sigma$ is the same: this is the Sharpe ratio for all efficient portfolios in the market with a risk-free asset, in particular, for the tangency portfolio. According to the result obtained in (b), this Sharpe ratio is greater than the Sharpe ratio for any other efficient portfolio for the market containing only risky assets. Consequently, the straight line (the efficient frontier of the market including the risk-free asset) is tangent to the hyperbola depicting the efficient frontier of the market with N risky assets, the point of tangency being (σ_v, m_v).

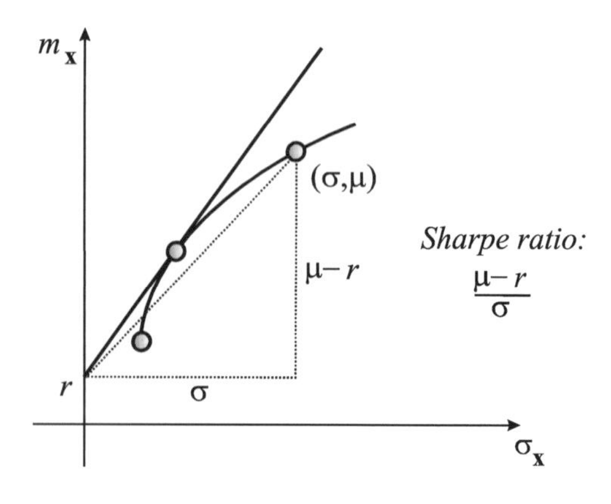

Fig. 10.1 Question 10.1: efficient frontiers and the Sharpe ratio

Question 10.2 (Problem ($\mathbf{M}^\mu$) for the Model Without a Risk-Free Asset) For the Markowitz model of a market of risky assets, consider the following portfolio selection problem. Minimize the variance $VarR_x$ of the return R_x on the portfolio $x = (x_1, \ldots, x_N)$ with $x_1 + \ldots + x_N = 1$ such that $ER_x \geq \mu$ where μ is some given level of the expected return.

Formally, given μ, consider the minimization problem:

$$(\mathbf{M}^\mu) \qquad \min_{x \in R^N} \sigma_x^2$$

subject to

$$x_1 + \ldots + x_N = 1$$

and

$$ER_x \geq \mu. \tag{10.5}$$

Let Assumptions 1 and 2 (see Chap. 2) hold.

(a) Show the following:
  (i) If μ is greater than the expected return on the minimum variance portfolio x^{MIN}, then the solution to problem ($\mathbf{M}^\mu$) is the portfolio

$$x^*(\mu) = g + h\mu, \tag{10.6}$$

  where

$$g = \frac{1}{D}[B\ We - A\ Wm], \tag{10.7}$$

$$h = \frac{1}{D}[C\ Wm - A\ We]. \tag{10.8}$$

  (ii) If μ is not greater than the expected return on the minimum variance portfolio x^{MIN}, then x^{MIN} is the solution to ($\mathbf{M}^\mu$).
(b) Show that g is a normalized portfolio and h is a self-financing portfolio.
(c) Compute the variance of the return on the optimal portfolio in problem ($\mathbf{M}^\mu$).

Answer

(a) For each $\tau \geq 0$, let x_τ^* denote the solution to the Markowitz optimization problem

$$(\mathbf{M}_\tau) \qquad \max_{x \in R^N} \{\tau ER_x - VarR_x\}$$

subject to

$$\langle x, e \rangle = 1.$$

If $\tau = 0$, then this problem reduces to

$$(\mathbf{M}_0) \qquad \min_{x \in R^N} VarR_x$$

subject to

$$\langle x, e \rangle = 1.$$

The solution to $(\mathbf{M}_0)$ is the minimum variance portfolio $x^{MIN} = x_0^*$. Let m_0^* denote its expected return. We have shown in Chap. 3 (Remark 3.5) that the expected return $m_\tau^* = ER_{x_\tau^*}$ on the portfolio x_τ^* is given by

$$ER_{x_\tau^*} = \frac{A}{C} + \frac{\tau}{2}\frac{D}{C} \ (\tau \geq 0). \tag{10.9}$$

By substituting $\tau = 0$ into the this formula, we obtain that the expected return on the minimum variance portfolio $x^{MIN} = x_0^*$ is

$$m_0^* = ER_{x_0^*} = \frac{A}{C}. \tag{10.10}$$

Consider two cases: (i) $\mu > m_0^*$ and (ii) $\mu \leq m_0^*$. In the latter case, the portfolio x^{MIN} satisfies constraint (10.5) because the expected return m_0^* on this portfolio is not less then μ. Therefore this portfolio, minimizing the variance $VarR_x$ among *all* normalized portfolios x, is the solution to problem $(\mathbf{M}^\mu)$. Thus, if $\mu \leq m_0^*$, then the constraint $ER_x \geq \mu$ in the optimization problem $(\mathbf{M}^\mu)$ is not binding.

Let us now assume that $\mu > m_0^*$ (case (i)). Since $m_0^* = \frac{A}{C}$ (see (10.10)) and $\mu > \frac{A}{C}$, there is a strictly positive number τ such that

$$\mu = \frac{A}{C} + \frac{\tau}{2}\frac{D}{C} \tag{10.11}$$

(recall that D and C are strictly positive). Consider the efficient portfolio x_τ^* corresponding to the risk tolerance τ defined by formula (10.11). We claim that this portfolio x_τ^* is the solution to optimization problem $(\mathbf{M}^\mu)$.

Suppose the contrary: there is a normalized portfolio x with $ER_x \geq \mu$ and $VarR_x < VarR_{x_\tau^*}$. But $\mu = ER_{x_\tau^*}$ by virtue of the definition of τ (see (10.11)) and by virtue of formula (10.9) for the expected return on the portfolio x_τ^*. Consequently,

$$ER_x \geq ER_{x_\tau^*}, \ VarR_x < VarR_{x_\tau^*}.$$

This implies

$$\tau ER_x - VarR_x > \tau ER_{x_\tau^*} - VarR_{x_\tau^*},$$

which means that x_τ^* is not a solution to the optimization problem ($\mathbf{M}_\tau$). A contradiction.

To compute the portfolio $x^*(\mu)$ solving the optimization problem ($\mathbf{M}^\mu$) in the case (i) it remains to express τ through μ by using formula (10.11) and substitute this value of τ into the expression for x_τ^*:

$$x_\tau^* = x^{MIN} + \frac{\tau}{2} z^* = \frac{We}{C} + \frac{\tau}{2}\left(Wm - \frac{A}{C}We\right) \tag{10.12}$$

(see Theorem 3.1 in Chap. 3). From (10.11), we find

$$\frac{\tau}{2} = \frac{C}{D}\left(\mu - \frac{A}{C}\right) = \frac{C}{D}\mu - \frac{A}{D}.$$

By substituting this value of $\frac{\tau}{2}$ into (10.12), we get

$$x^*(\mu) = x_\tau^* = \frac{We}{C} + \left(\frac{C}{D}\mu - \frac{A}{D}\right)\left(Wm - \frac{A}{C}We\right)$$

$$= \frac{We}{C} - \frac{A}{D}(Wm - \frac{A}{C}We) \tag{10.13}$$

$$+ \mu\frac{C}{D}(Wm - \frac{A}{C}We). \tag{10.14}$$

The expression in (10.13) can be transformed to

$$\frac{We}{C} - \frac{A}{D}\left(Wm - \frac{A}{C}We\right) = -\frac{A}{D}Wm + \frac{D + A^2}{DC}We$$
$$= -\frac{A}{D}Wm + \frac{B}{D}We = g$$

(recall that $D = BC - A^2$), and the expression in (10.14) can be written as

$$\mu\frac{C}{D}(Wm - \frac{A}{C}We) = \mu\frac{1}{D}(C\,Wm - A\,We) = \mu h.$$

This proves formula (10.6).

(b) Let us show that g is a normalized portfolio and h is a self-financing portfolio. This follows from the computations:

$$\langle e, g\rangle = \frac{1}{D}\left[B\,\langle e, We\rangle - A\,\langle e, Wm\rangle\right] = \frac{1}{D}[B\,C - A^2] = 1$$

and

$$\langle e, h\rangle = \frac{1}{D}[C\,\langle e, Wm\rangle - A\,\langle e, We\rangle] = \frac{1}{D}[C\,A - A\,C] = 0.$$

(Recall that $A = \langle e, Wm\rangle$, $B = \langle m, Wm\rangle$, $C = \langle e, We\rangle$, $D = BC - A^2$.)

(c) To compute the variance of the return on the optimal portfolio in problem $(\mathbf{M}^{\mu})$, we use the formulas obtained in Chap. 4:

$$ER_{x_\tau^*} = \frac{A}{C} + \frac{\tau}{2}\frac{D}{C}, \tag{10.15}$$

$$VarR_{x_\tau^*} = \frac{1}{C} + \frac{\tau^2}{4}\frac{D}{C}. \tag{10.16}$$

If $\mu \le m_0^*$, then, as has been proved, the solution to $(\mathbf{M}^{\mu})$ is the minimum variance portfolio $x^{MIN} = x_0^*$. The variance of the return on this portfolio is equal to $\frac{1}{C}$ (substitute $\tau = 0$ into (10.16)).

Let $\mu > m_0^*$. Then $x^*(\mu) = x_\tau^*$, where $\mu = \frac{A}{C} + \frac{\tau}{2}\frac{D}{C}$. Therefore $\frac{\tau}{2} = \frac{C}{D}\mu - \frac{A}{D}$ (we have already used this equality), and so

$$VarR_{x^*(\mu)} = VarR_{x_\tau^*} = \frac{1}{C} + (\frac{C}{D}\mu - \frac{A}{D})^2\frac{D}{C}. \tag{10.17}$$

To simplify the latter expression, we write

$$\frac{1}{C} + (\frac{C}{D}\mu - \frac{A}{D})^2\frac{D}{C} = \frac{1}{C} + \frac{C}{D}\mu^2 - 2\mu\frac{A}{D} + \frac{A^2}{DC}$$
$$= \frac{1}{D}(C\mu^2 - 2A\mu + \frac{D + A^2}{C}) = \frac{1}{D}(C\mu^2 - 2A\mu + B).$$

Thus, if $\mu > m_0^*$, the answer is

$$VarR_{x^*(\mu)} = \frac{1}{D}(C\mu^2 - 2A\mu + B). \tag{10.18}$$

Question 10.3 (Problem ($\mathbf{M}^\mu$) for the Model with a Risk-Free Asset)

(a) Assuming that the market contains a risk-free asset with non-random return $r > 0$, find the solution $\mathbf{x}^*(\mu)$ to the following problem:

($\mathbf{M}^\mu$) Minimize the variance $Var\mathbf{R}_\mathbf{x}$ of the return $\mathbf{R}_\mathbf{x}$ on the portfolio $\mathbf{x} = (x_0, x_1, \ldots, x_N) \in R^N$ under the constraints

$$E\mathbf{R}_\mathbf{x} \geq \mu,$$

$$\sum_{i=0}^{N} x_i = 1,$$

where μ is some given level of expected return.

(b) According to Tobin's mutual fund theorem, every efficient portfolio—in particular, $\mathbf{x}^*(\mu)$—is a combination $(1-\gamma)\mathbf{x}^{MIN} + \gamma\mathbf{v}$ of the minimum variance portfolio $\mathbf{x}^{MIN} = (1, 0, \ldots, 0)$ and the tangency portfolio $\mathbf{v} = \mathbf{x}^*_{\bar{\tau}}$ with weights $1-\gamma$ and γ. Show that the value of $\gamma = \gamma(\mu)$ for which

$$\mathbf{x}^*(\mu) = (1-\gamma(\mu))\mathbf{x}^{MIN} + \gamma(\mu)\mathbf{v}$$

is given by the formula

$$\gamma(\mu) = \frac{(\mu - r)(A - rC)}{H}$$

if $\mu > r$, and $\gamma(\mu) = 0$ if $\mu \leq r$.

Answer

(a) If $\mu \leq r$, then the solution to the problem ($\mathbf{M}^\mu$) is the portfolio $\mathbf{e}_0 = (1, 0, \ldots, 0)$. Indeed, the return on $\mathbf{e}_0$ is r (and so the constraint $E\mathbf{R}_\mathbf{x} \geq \mu$ is satisfied), while the variance of return on this portfolio is zero, which an absolute minimum for a variance.

Let us assume that $\mu > r$. In Chap. 6 (Eq. (6.1)), it was proved that the expected return on the $\mathbf{x}^*_\tau$ efficient portfolio $\mathbf{x}^*_\tau$ corresponding to the risk tolerance τ, is given by the formula

$$E\mathbf{R}_{\mathbf{x}^*_\tau} = r + \frac{\tau}{2}H.$$

Consider that value of $\tau = \tau(\mu)$ for which

$$\mu = r + \frac{\tau}{2}H,$$

in other words, set

$$\tau = \tau(\mu) = 2 \cdot \frac{\mu - r}{H}.$$

We have $\tau(\mu) > 0$ because $\mu > r$. Consider the efficient portfolio $\mathbf{x}_\tau^*$ corresponding to the risk tolerance $\tau = \tau(\mu)$. We claim that this portfolio is the solution to optimization problem $(\mathbf{M}^\mu)$.

Suppose the contrary: there is a normalized portfolio $\mathbf{x}$ with $E\mathbf{R}_\mathbf{x} \geq \mu$ and $Var\mathbf{R}_{\mathbf{x}_\tau} < Var\mathbf{R}_{\mathbf{x}_\tau^*}$. But $\mu = E\mathbf{R}_{\mathbf{x}_\tau^*}$ by virtue of the definition of $\tau = \tau(\mu)$ and the formula $\mu = r + \frac{\tau}{2}H$ for the expected return of the portfolio $\mathbf{x}_\tau^*$. Consequently,

$$E\mathbf{R}_\mathbf{x} \geq E\mathbf{R}_{\mathbf{x}_\tau^*},$$

and

$$Var\mathbf{R}_\mathbf{x} < Var\mathbf{R}_{\mathbf{x}_\tau^*}.$$

This implies

$$\tau E\mathbf{R}_\mathbf{x} - Var\mathbf{R}_\mathbf{x} > \tau E\mathbf{R}_{\mathbf{x}_\tau^*} - Var\mathbf{R}_{\mathbf{x}_\tau^*},$$

which means that $\mathbf{x}_\tau^*$ is not a solution to the optimization problem $(\mathbf{M}_\tau)$. A contradiction.

To compute the portfolio $\mathbf{x}^*(\mu)$ solving the optimization problem $(\mathbf{M}^\mu)$ it remains to substitute $\tau = \tau(\mu) = 2 \cdot \frac{\mu - r}{H}$ into the expression for $\mathbf{x}_\tau^*$ obtained in Chap. 5:

$$\left(1 + \frac{\tau}{2}(rC - A), \frac{\tau}{2}W(m - re)\right) = \mathbf{x}_\tau^*$$
$$= \left(1 + \frac{(\mu - r)}{H}(rC - A), \frac{(\mu - r)W(m - re)}{H}\right).$$

Thus *the solution to the Markowitz problem* $(\mathbf{M}^\mu)$ *for the market with a risk-free asset is given by*

$$\mathbf{x}^*(\mu) = \left(1 + \frac{(\mu - r)(rC - A)}{H}, \frac{(\mu - r)W(m - re)}{H}\right).$$

(b) To find $\gamma = \gamma(\mu)$ for $\mu > r$, we write

$$\mathbf{x}^*(\mu) = (1 - \gamma)\mathbf{x}^{MIN} + \gamma\mathbf{v}.$$

Since

$$\mathbf{v} = \frac{W(m - re)}{A - rC},$$

we obtain

$$\gamma \frac{W(m - re)}{A - rC} = \frac{(\mu - r)W(m - re)}{H},$$

which yields

$$\gamma(\mu) = \frac{(\mu - r)(A - rC)}{H}.$$

Question 10.4 (A Numerical Example on Portfolio Selection) There are two risky assets $i = 1, 2$ with random returns R_1 and R_2. The expectations of the returns $m_i = ER_i$ and their covariances $\sigma_{ij} = Cov(R_i, R_j)$ are given by:

$$m_1 = 1, \; m_2 = 2;$$

$$\sigma_{11} = 3, \; \sigma_{12} = \sigma_{21} = 3, \; \sigma_{22} = 5.$$

(a) For each $\tau \geq 0$, find the mean-variance efficient portfolio x_τ^* of an investor with risk tolerance τ.

[When answering this question, you can use the formulas derived in Chap. 2 and the following formula for the inverse matrix V^{-1}. If

$$V = \begin{pmatrix} \sigma_{11} & \sigma_{12} \\ \sigma_{21} & \sigma_{22} \end{pmatrix},$$

then

$$V^{-1} = \begin{pmatrix} \dfrac{\sigma_{22}}{|V|} & \dfrac{-\sigma_{12}}{|V|} \\ \dfrac{-\sigma_{21}}{|V|} & \dfrac{\sigma_{11}}{|V|} \end{pmatrix},$$

where $|V| = \sigma_{11}\sigma_{22} - \sigma_{12}\sigma_{21}$ is the determinant of the matrix V.]

(b) Compute the expectation $\mu(\tau) = ER_{x_\tau^*}$ of the return on the portfolio x_τ^* and its variance $\sigma^2(\tau) = VarR_{x_\tau^*}$ by using the formulas

$$ER_x = \langle m, x \rangle, \; VarR_x = \langle x, Vx \rangle,$$

which were given in the chapter.

(c) For the market with the two risky assets, draw the diagram depicting the efficient frontier in the σ_x–m_x plane. By using (b), derive the equation of the curve representing the efficient frontier. Indicate that point on the curve which corresponds to the minimum variance portfolio. What are the coordinates of this point?

(d) Suppose there is a risk-free asset whose return r is non-random: $r = 0.1$. On the same diagram as constructed in (c) draw, additionally, the efficient frontier for the market with three assets, one risk-free $i = 0$ and two risky $i = 1, 2$. Indicate on this diagram the point $(\bar{\sigma}, \bar{\mu})$ corresponding to the tangency portfolio. (Here, $\bar{\sigma} = \sigma(\bar{\tau})$ is the standard deviation and $\bar{\mu} = \mu(\bar{\tau})$ is the expectation of the return on the tangency portfolio.)

(e) By using (b), show that in the example at hand the tangency portfolio $x^*_{\bar{\tau}}$ corresponds to the risk tolerance $\bar{\tau} = 20/3$, and its Sharpe ratio is equal to $\sqrt{0.77}$.

Answer

(a) We use the formula

$$x^*_\tau = x^{MIN} + \frac{\tau}{2} z^*,$$

where

$$x^{MIN} = \frac{We}{\langle e, We \rangle} \ (= x^*_0), \ e = (1, 1),$$

and

$$z^* = Wm - \frac{\langle e, Wm \rangle}{\langle e, We \rangle} We.$$

The covariance matrix V is as follows:

$$V = \begin{pmatrix} 3 & 3 \\ 3 & 5 \end{pmatrix}.$$

We have $|V| = 3 \cdot 5 - 3 \cdot 3 = 15 - 9 = 6$, and so

$$W = V^{-1} = \begin{pmatrix} \sigma_{22}/|V| & -\sigma_{12}/|V| \\ -\sigma_{21}/|V| & \sigma_{11}/|V| \end{pmatrix} = \begin{pmatrix} 5/6 & -1/2 \\ -1/2 & 1/2 \end{pmatrix}.$$

Consequently,

$$We = \begin{pmatrix} 5/6 & -1/2 \\ -1/2 & 1/2 \end{pmatrix} \begin{pmatrix} 1 \\ 1 \end{pmatrix} =$$

$$\begin{pmatrix} 5/6 - 1/2 \\ -1/2 + 1/2 \end{pmatrix} = \begin{pmatrix} 1/3 \\ 0 \end{pmatrix},$$

$$\langle e, We \rangle = 1 \cdot 1/3 + 1 \cdot 0 = 1/3,$$

and

$$x^{MIN} = \frac{We}{\langle e, We \rangle} = \frac{(1/3, 0)}{1/3} = (1, 0).$$

To compute z^*, we write

$$Wm = \begin{pmatrix} 5/6 & -1/2 \\ -1/2 & 1/2 \end{pmatrix} \begin{pmatrix} 1 \\ 2 \end{pmatrix} =$$

$$\begin{pmatrix} 5/6 - (1/2) \cdot 2 \\ -1/2 + (1/2) \cdot 2 \end{pmatrix} = \begin{pmatrix} -1/6 \\ 1/2 \end{pmatrix},$$

$$\langle e, Wm \rangle = 1 \cdot (-1/6) + 1 \cdot (1/2) = 1/3.$$

Thus, we obtain

$$z^* = Wm - \frac{\langle e, Wm \rangle}{\langle e, We \rangle} We = (-1/6, 1/2) - \frac{1/3}{1/3}(1/3, 0) =$$

$$(-1/6, 1/2) - (1/3, 0) = (-1/2, 1/2),$$

and so $x^*_\tau = x^{MIN} + \frac{\tau}{2} z^* = (1, 0) + \frac{\tau}{2}(-1/2, 1/2) = (1 - \tau/4, \tau/4)$.

(b) By using these formulas, we get

$$ER_{x^*_\tau} = \langle m, x^*_\tau \rangle = 1 \cdot (1 - \frac{\tau}{4}) + 2 \cdot (\frac{\tau}{4}) =$$

$$1 - \frac{\tau}{4} + \frac{\tau}{2} = 1 + \frac{\tau}{4};$$

$$Vx^*_\tau = \begin{pmatrix} 3 & 3 \\ 3 & 5 \end{pmatrix} \begin{pmatrix} 1 - \frac{\tau}{4} \\ \frac{\tau}{4} \end{pmatrix} =$$

$$\begin{pmatrix} 3 \cdot (1 - \frac{\tau}{4}) + 3 \cdot (\frac{\tau}{4}) \\ 3 \cdot (1 - \frac{\tau}{4}) + 5 \cdot \frac{\tau}{4} \end{pmatrix} =$$

$$\begin{pmatrix} 3 \\ 3 + \frac{\tau}{2} \end{pmatrix},$$

which yields

$$\langle x^*_\tau, Vx^*_\tau \rangle = (1 - \frac{\tau}{4}) \cdot 3 + (\frac{\tau}{4}) \cdot (3 + \frac{\tau}{2}) =$$

$$3 - 3\frac{\tau}{4} + 3\frac{\tau}{4} + \frac{\tau^2}{8} = 3 + \frac{\tau^2}{8}.$$

Thus

$$\mu(\tau) = 1 + \frac{\tau}{4}, \; \sigma^2(\tau) = 3 + \frac{\tau^2}{8}.$$

(c) By using the two equations

$$\mu(\tau) = 1 + \frac{\tau}{4}, \; \sigma^2(\tau) = 3 + \frac{\tau^2}{8},$$

we can express σ^2 through μ. From the first equation, we get

$$\tau = 4(\mu - 1),$$

and by substituting this expression for τ in the second equation, we find

$$\sigma^2 = 3 + \frac{\tau^2}{8} = 3 + 4(\mu - 1)^2/2 =$$

$$3 + 2(\mu - 1)^2 = 2\mu^2 - 4\mu + 5,$$

and so

$$\sigma^2 = 2\mu^2 - 4\mu + 5.$$

The equation of the curve representing the efficient frontier in the σ_x-μ_x plane is as follows:

$$\sigma = \sqrt{2\mu^2 - 4\mu + 5}.$$

The graph of this function is a hyperbola. The efficient frontier in the σ_x–μ_x plane is the upper part of this hyperbola, see Fig. 10.2.

The coordinates of the point corresponding to the minimum variance portfolio are $\sigma = \sqrt{3}$ and $\mu = 1$.

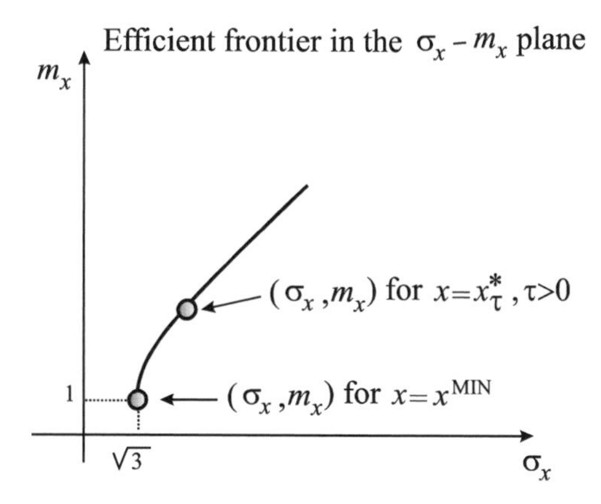

Fig. 10.2 Question 10.4: efficient frontier

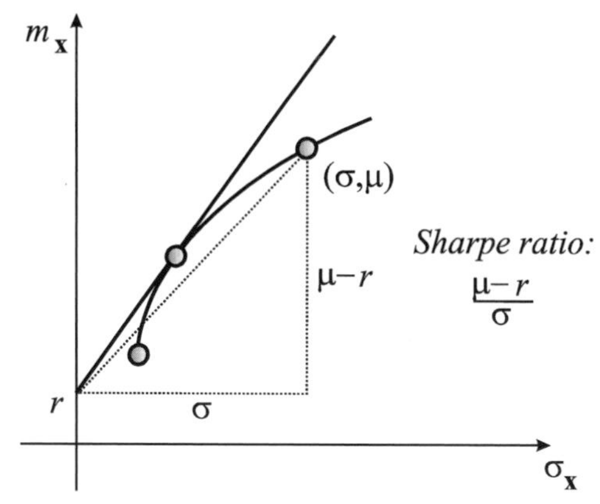

Fig. 10.3 Question 10.4: tangency portfolio

(d) The efficient frontier for the market with three assets, one risk-free $i = 0$ and two risky $i = 1, 2$, is the straight line (ray) emanating from the point $(0, r)$ and tangent to the hyperbola representing the efficient frontier of the market with two assets $i = 1, 2$. The point $(\bar{\sigma}, \bar{\mu})$ of tangency of the straight line and the hyperbola corresponds to the tangency portfolio, see Fig. 10.3.

(Here $\bar{\sigma} = \sigma(\bar{\tau})$ is the standard deviation and $\bar{\mu} = \mu(\bar{\tau})$ is the expectation of the return on the tangency portfolio.)

(e) In the example at hand, the Sharpe ratio of an efficient portfolio x^*_τ is given by

$$\rho(\tau) = \frac{\mu(\tau) - r}{\sigma(\tau)} = \frac{0.9 + \frac{\tau}{4}}{\sqrt{3 + \frac{\tau^2}{8}}},$$

because, as we have obtained in (b),

$$\mu(\tau) = 1 + \frac{\tau}{4}, \ \sigma^2(\tau) = 3 + \frac{\tau^2}{8}.$$

The tangency portfolio corresponds to the level of risk tolerance

$$\bar{\tau} = \frac{2}{A - rC},$$

where

$$A = \langle e, Wm \rangle = 1/3,$$
$$C = \langle e, We \rangle = 1/3$$

(see (a)). Thus

$$\bar{\tau} = \frac{2}{1/3 - 0.1 \cdot (1/3)} = \frac{2}{9/30} = \frac{20}{3}.$$

We have shown that

$$\rho(\tau) = \frac{\mu(\tau) - r}{\sigma(\tau)} = \frac{0.9 + \frac{\tau}{4}}{\sqrt{3 + \frac{\tau^2}{8}}}.$$

By substituting $\bar{\tau} = 20/3$, we find

$$\rho(\bar{\tau}) = \frac{0.9 + 10/(3 \cdot 2)}{\sqrt{3 + 100/(9 \cdot 2)}} =$$

$$\frac{0.9 + 5/3}{\sqrt{3 + 50/9}} = \frac{2.7 + 5}{\sqrt{27 + 50}} = \frac{7.7}{\sqrt{77}} = \sqrt{0.77}.$$

Question 10.5 (Jensen's Index and Sharpe Ratio) The mutual fund F has the 10-year record of rates of return $R^F(t)$ ($t = 1, 2, \ldots, 10$) shown in the second column in the table below (Table 10.1). The rates of return on the market portfolio, $R^M(t)$, and on cash, $r(t)$, during the 10-year period are indicated in the third and in the fourth columns of the table.

(a) Compute the averages of $R^F(t)$, $R^M(t)$ and $r(t)$ over the 10-year period. Using the first and the second of these averages as proxies for the expected values $m^F = ER^F$ and $m^M = ER^M$, respectively, estimate the variances $Var(R^M)$, $Var(R^F)$, the standard deviations $\sigma^M = \sqrt{Var(R^M)}$, $\sigma^F = \sqrt{Var(R^F)}$ and the covariance $Cov(R^M, R^F)$. Compute the beta of the mutual fund F.

Table 10.1 Annual returns of the mutual fund, market portfolio and cash over 10 years

Rates of return			
Year	R^F	R^M	r
1	0.2	0.2	0.09
2	−0.1	−0.1	0.08
3	0.1	0.2	0.08
4	0.2	0.1	0.08
5	0.2	0.1	0.08
6	0.1	0.1	0.07
7	0.2	0.2	0.08
8	0.1	0.2	0.07
9	−0.2	−0.1	0.08
10	0.2	0.1	0.09

(b) Suppose that cash is a risk-free asset whose rate of return r is equal to the average of $r(t)$ $(t = 1, 2, \ldots, 10)$ calculated in (a) and assume that the market portfolio is efficient. Draw the diagram of the efficient frontier in the $\sigma_{\mathbf{x}}$–$m_{\mathbf{x}}$ plane. Indicate on the diagram the point corresponding to the portfolio containing only the risk-free asset and the point corresponding to the market portfolio. Compute the coordinates of these points. Calculate the Sharpe ratio ρ^M of the market portfolio.

(c) Evaluate the performance of the mutual fund F by comparing the expected returns of this fund with those predicted by the CAPM: calculate the Jensen index of the fund F and find out whether it is positive, negative or equal to zero. Check whether the fund F has a sufficiently high level of efficiency—in the sense that the Sharpe ratio ρ^M of the market portfolio does not exceed by more than 10 % the Sharpe ratio ρ^F of the fund F (check whether $(\rho^M - \rho^F)/\rho^M \le 0.1$).

[The variance of a random variable R can be estimated by using the formula

$$VarR = \frac{1}{n-1} \sum_{t=1}^{n} (R(t) - ER)^2,$$

where ER is the average of the observed values $R(t)$. The formula for estimating the covariance between two random variables R_1 and R_2 based on the observed values $R_1(1), \ldots, R_1(n)$ and $R_2(1), \ldots, R_2(n)$ is as follows:

$$Cov(R_1, R_2) = \frac{1}{n-1} \sum_{t=1}^{n} (R_1(t) - ER_1)(R_2(t) - ER_2),$$

where ER_1 and ER_2 are the averages of $R_1(t)$ and $R_2(t)$, respectively.]

Table 10.2 Question 10.5: calculations of empirical variances and covariances

$Er = 0.08,\ m^F = ER^F = 0.1,\ m^M = ER^M = 0.1,$					
r	R^F	R^M	$(R^F - ER^F)^2$	$(R^M - m^M)^2$	$(R^M - m^M)(R^F - m^F)$
0.09	0.2	0.2	0.01	0.01	0.01
0.08	−0.1	−0.1	0.04	0.04	0.04
0.08	0.1	0.2	0	0.01	0
0.08	0.2	0.1	0.01	0	0
0.08	0.2	0.1	0.01	0	0
0.07	0.1	0.1	0	0	0
0.08	0.2	0.2	0.01	0.01	0.01
0.07	0.1	0.2	0	0.01	0
0.08	−0.2	−0.1	0.09	0.04	0.06
0.09	0.2	0.1	0.01	0	0

Answer (a) We compute the averages m^F, m^M, the variances $Var(R^M)$, $Var(R^F)$, the standard deviations $\sigma^M = \sqrt{Var(R^M)}$, $\sigma^F = \sqrt{Var(R^F)}$, and the covariance $Cov(R^M, R^F)$ (Table 10.2):

Consequently

$$VarR^F = \frac{\sum(R^F - ER^F)^2}{9} = 0.02,\ \sigma^F = \sqrt{0.02} = 0.141,$$

$$VarR^M = \frac{\sum(R^M - ER^M)^2}{9} = 0.0133,$$
$$\sigma^M = \sqrt{0.0133} = 0.115,$$

and

$$Cov(R^M, R^F) = \frac{\sum(R^M - ER^M)(R^F - ER^F)}{9} = 0.0133.$$

We obtain the value of the beta of the fund F:

$$\beta^F = \frac{Cov(R^M, R^F)}{Var(R^M)} = 1.$$

(b) The efficient frontier in the $\sigma_{\mathbf{x}}$–$m_{\mathbf{x}}$ plane for a market with a risk-free asset is the straight line containing the point $(0, r)$ and the point (σ^M, m^M), where $m^M = ER^M$ is the expected return on the market portfolio and σ^M is the standard deviation of this return. By using the results obtained in (a), we find

$$(0, r) = (0, 0.08),\ (\sigma^M, m^M) = (0.115, 0.1).$$

The Sharpe ratio ρ^M of the market portfolio is

$$\rho^M = \frac{m^M - r}{\sigma^M} = \frac{0.02}{0.115} = 0.174.$$

The number

$$\rho^M = \frac{m^M - r}{\sigma^M} = 0.174$$

is the slope of the straight line representing the efficient frontier, as shown in Fig. 10.4.

(c) The Jensen index of the fund F is

$$J^F = ER^F - [r + \beta^F(ER^M - r)] = 0.1 - 0.1 = 0.$$

Thus the expected returns of the fund are exactly equal to those predicted by the CAPM.

To estimate the efficiency of F, we compute its Sharpe ratio ρ^F

$$\rho^F = \frac{m^F - r}{\sigma^F} = \frac{0.1 - 0.08}{\sigma^F} = \frac{0.02}{0.141} = 0.141.$$

We have found above that the Sharpe ratio ρ^M of the market portfolio is

$$\rho^M = \frac{m^M - r}{\sigma^M} = 0.174.$$

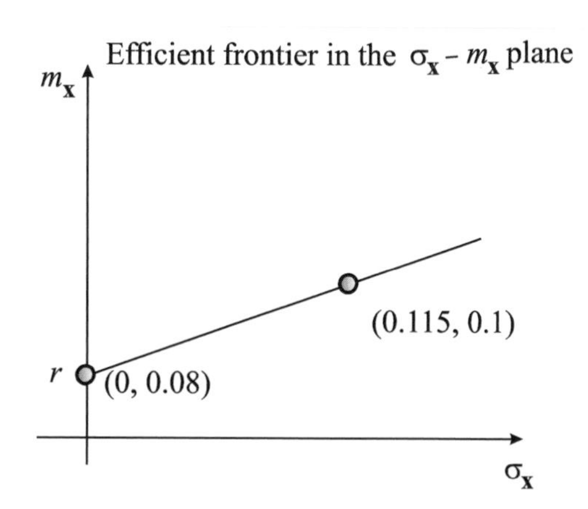

Fig. 10.4 Question 10.5: efficient frontier

Since the efficient frontier is the straight line, we conclude that $\rho^M = 0.174$ is the Sharpe ratio for all efficient portfolios. Finally, we have

$$\frac{\rho^M - \rho^F}{\rho^M} = \frac{0.174 - 0.141}{0.174} = \frac{0.033}{0.174} = 0.19 > 10\,\%,$$

and so the fund does not have a sufficiently high level of efficiency.

Part II

Derivative Securities Pricing

11 Dynamic Securities Market Model

11.1 Multi-Period Model of an Asset Market

The Basic Elements of the Model

- There are $T + 1$ dates: $t = 0, \ldots, T$. At any of these dates, trading on the securities market is possible.
- A finite set

$$A = \{a^1, \ldots, a^L\}$$

is given, elements of which are interpreted as possible *states of the world*. At each time $t = 1, 2, \ldots, T$, any of these states of the world can be realized. That state which is realized at time t is denoted by a_t. By means of a_t we represent the totality of all economic and financial factors that influence the market at time t.
- The sequence

$$\omega = (a_1, \ldots, a_T)$$

is called the *history* of the market over the time period $1, 2, \ldots, T$. Such sequences $\omega = (a_1, \ldots, a_T)$ may also be viewed as possible *scenarios* of the market development over the time period $1 \leq t \leq T$. For each $t = 1, 2, \ldots, T - 1$, the sequence

$$\omega^t = (a_1, \ldots, a_t)$$

is called the *partial history* or *partial scenario* (up to time t). There are L^T histories and L^t partial histories for each $t < T$.

I.V. Evstigneev et al., *Mathematical Financial Economics*, Springer Texts in Business and Economics, DOI 10.1007/978-3-319-16571-4_11

- At time 0, the future realization of $\omega = (a_1, \ldots, a_T)$ is not known. All what is known is the *probability distribution* P on the set of all possible scenarios ω. For each

$$\omega = (a_1, \ldots, a_T),$$

we are given a non-negative number $P(\omega) \geq 0$: the probability that the market will develop according to the scenario ω. The sum of all these probabilities is equal to 1:

$$\sum_{\omega} P(\omega) = 1.$$

Since there are L^T possible scenarios (L being the number of possible states of the world at each time t), there are L^T summands in the above sum.

- There are $N + 1$ securities (assets) $i = 0, 1, \ldots, N$. The vector of their prices at time $t = 0, \ldots, T$ is denoted by

$$\mathbf{S}_t = (S_t^0, S_t^1, \ldots, S_t^N).$$

It is supposed that the prices S_t^i $(t \geq 1)$ of the securities depend on the current situation a_t and on the factors $a_1, \ldots, a_{t-1}$ that influenced the market prior to time t. This is expressed by the assumption that the vector $\mathbf{S}_t$ $(t \geq 1)$ is a *function* of the sequence $\omega^t = (a_1, \ldots, a_t)$ of the states of the world up to time t:

$$\mathbf{S}_t = \mathbf{S}_t(a_1, \ldots, a_t).$$

In other words, every coordinate S_t^i of the vector $\mathbf{S}_t$ (the price of asset i at time $t \geq 1$) is the given function

$$S_t^i = S_t^i(a_1, \ldots, a_t)$$

of the sequence $\omega^t = (a_1, \ldots, a_t)$.

- The 0th asset $i = 0$ plays a special role. Its price S_t^0 at time t (i.e., the 0th coordinate of the vector $\mathbf{S}_t$) is assumed to be non-random and equal to $(1 + r)^t$, where $r > 0$ is some given number. Asset $i = 0$ may be interpreted as cash, with r being the interest rate (the same for lending and borrowing). The vector $\mathbf{S}_0 = (S_0^0, S_0^1, \ldots, S_0^N)$ is *constant* (non-random).

Feasible Scenarios/Histories Denote by Ω the set of all market scenarios ω that have strictly positive probability: $P(\omega) > 0$ (*feasible* scenarios). They are of primary interest, since all the others, occurring with zero probability, can essentially be excluded from consideration.

For a random variable $X(\omega)$, we denote by EX its expectation with respect to the probability measure[1] P:

$$EX = \sum_{\omega} P(\omega)X(\omega).$$

where the sum is taken over all ω. Equivalently, this sum can be restricted to feasible scenarios:

$$EX = \sum_{\omega \in \Omega} P(\omega)X(\omega),$$

since for all the others $P(\omega) = 0$.

Investor's Portfolios A *portfolio* that can be selected by an investor *at time* 0,

$$\mathbf{h}_0 = (h_0^0, \ldots, h_0^N),$$

is a vector in R^{N+1} whose coordinate h_0^i indicates the number of ("physical") units of asset i held in the portfolio. The coordinates can be either positive or negative. The latter possibility means that *short sales* are allowed.[2]

It is assumed that the investor can observe the market history $\omega^t = (a_1, a_2, \ldots, a_t)$ and use this information for constructing a portfolio at time t. Thus, the investor's portfolio—contingent on the observed history $\omega^t = (a_1, a_2, \ldots, a_t)$—is a vector function

$$\mathbf{h}_t = \mathbf{h}_t(\omega^t) = \mathbf{h}_t(a_1, \ldots, a_t),$$

where the coordinates h_t^i of the vector $\mathbf{h}_t = (h_t^0, \ldots, h_t^N)$ are the portfolio positions.

The value of the portfolio $\mathbf{h}_t = (h_t^0, h_t^1, \ldots, h_t^N)$ at time t is given by the scalar product

$$\langle \mathbf{S}_t, \mathbf{h}_t \rangle = S_t^0 h_t^0 + S_t^1 h_t^1 + \ldots + S_t^N h_t^N.$$

Here, the prices $S_t^i = S_t^i(\omega^t)$ and the portfolio positions $h_t^i = h_t^i(\omega^t)$ depend on the market history $\omega^t = (a_1, \ldots, a_t)$, and so does $\langle \mathbf{S}_t, \mathbf{h}_t \rangle$.

Trading Strategies A *trading strategy* $\mathbf{H}$ is a sequence

$$\mathbf{H} = (\mathbf{h}_0, \ldots, \mathbf{h}_T),$$

[1]The terms "probability measure", "probability distribution", or simply "probability" are used interchangeably.

[2]In the mean-variance portfolio theory, we characterized portfolio positions, basically, in terms of the sums of money invested in one asset or another. Here, it is more convenient to deal with units of assets.

where $\mathbf{h}_0$ is the investor's portfolio at time 0, and

$$\mathbf{h}_t = \mathbf{h}_t(\omega^t),\ t = 1, 2, \dots, T,$$

are contingent portfolios which the investor is going to choose at times $t = 1, 2, \dots, T$ depending on the history $\omega^t = (a_1, \dots, a_t)$. By selecting a trading strategy, the investor specifies what portfolio he/she is going to have at *each* time $t = 0, 1, \dots, T$ in *each* possible contingency.

Self-Financing Trading Strategies The theory under consideration primarily focuses on those trading strategies $\mathbf{H} = (\mathbf{h}_0, \dots, \mathbf{h}_T)$ that satisfy

$$\langle \mathbf{S}_t, \mathbf{h}_{t-1} \rangle = \langle \mathbf{S}_t, \mathbf{h}_t \rangle,\ t = 1, 2, \dots, T. \tag{11.1}$$

Such strategies are called *self-financing*. Condition (11.1) means that the investor at each time period replaces the old portfolio $\mathbf{h}_{t-1}$ by the new one $\mathbf{h}_t$ without violating the budget constraint.

According to (11.1), the value $\langle \mathbf{S}_t, \mathbf{h}_t \rangle$ of the new portfolio $\mathbf{h}_t$ expressed in terms of the new price system $\mathbf{S}_t = (S_t^0, S_t^1, \dots, S_t^N)$ is equal to the value $\langle \mathbf{S}_t, \mathbf{h}_{t-1} \rangle$ of the old portfolio $\mathbf{h}_{t-1}$ expressed in terms of the price system $\mathbf{S}_t$. Such strategies exclude consumption and do not involve an inflow of external funds. During the time period $0 \leq t \leq T$, the investor "plays" on price changes with the view to obtaining, at the end of the period, a portfolio $\mathbf{h}_T$ that is most preferred in the sense of some (objective or subjective) criterion.

Fundamental Assumptions The model under consideration is based on the fundamental assumptions. The market we deal with is *frictionless*—assets can be sold and purchased in any quantities *without transaction costs*. There are *no constraints* on portfolios. Portfolio positions (components of the vectors $\mathbf{h}_t$) can be both positive and negative—*short sales are allowed*. Furthermore, it is supposed that there is an *ideal bank account* where one can deposit and borrow cash at the same *interest rate* r.

11.2 Basic Securities and Derivative Securities

Contingent Claims Any real-valued function $X(\omega)$ of the market history ω will be called a *contingent claim*. A contingent claim is interpreted as a contract that allows its owner to receive the specified amount of money $X(\omega)$ at time T. This amount might be both positive and negative; in the latter case the contract owner is obliged to pay, rather than receive, the specified amount. In general, $X(\omega)$ might depend on the whole market history $\omega = (a_1, a_2, \dots, a_T)$—on the current state of the world, a_T, at time T and on the previous states $a_1, a_2, \dots, a_{T-1}$.

Contingent Claims as Securities A contingent claim X can be regarded as a security (asset) which can, as well as the basic securities $i = 0, 1, 2, \dots, N$, be traded on the market. The value of this security at time T is equal to $X(\omega)$. The amount $X(\omega)$, and hence the value, might be negative if the contract involves the obligation to pay, rather than the right to receive, the contractually agreed sum of money.

Derivative Securities Important examples of contingent claims are derivative securities. Consider some of the basic assets $i = 1, 2, \dots, N$, say $i = 1$. For shortness, let us write S_t in place of S_t^1. A contingent claim of the form

$$X = F(S_0, S_1, \dots, S_T),$$

where F is some function of $T + 1$ variables, is called a *derivative security* for the *underlying security* $i = 1$. Since the prices $S_1, S_2, \dots, S_T$ are functions of the partial histories

$$\omega^1 = a_1,\ \omega^2 = (a_1, a_2), \dots,\ \omega^T = (a_1, \dots, a_T),$$

the contingent claim

$$X = X(\omega)$$

depends, generally, on the whole market history $\omega = \omega^T = (a_1, \dots, a_T)$. The term "derivative" points to the fact that the payoff $X(\omega)$—the price of the derivative security at time T—is completely determined by the prices $S_0, S_1, \dots, S_T$ of the underlying security.

The function F is called the *payoff function* of the derivative security. In many important cases, this function depends only on the last argument S_T:

$$X = F(S_T),$$

and so the price of the derivative security at time T can directly be computed as a function of the price S_T of the underlying security at time T.

European Options Typical examples of derivative securities with payoff of the form $X = F(S_T)$ are European call and put options. For a *European call*,

$$X^c = \max\{S_T - K, 0\}, \tag{11.2}$$

where $K > 0$ is the *strike price* (or the *exercise price*) of the option. This option gives the right, but not the obligation, to *buy* the underlying asset at time T at the fixed price K. Clearly, this right is used by the option owner only if $S_T > K$. In this case the option is worth $S_T - K$. If $S_T \leq K$, the option is useless, and so it is worth zero. Consequently, the price (or, equivalently, the payoff) of the option at time T is equal to the number X^c defined by formula (11.2).

Formally, the payoff function $F^c(S)$ of the European call option is

$$F^c(S) = \max\{S - K, 0\} = (S - K)^+.$$

The symbol S^+ is used as an abbreviation for $\max\{S, 0\}$.

Analogously, a *European put* option is defined as the derivative security whose payoff is

$$X^p = \max\{K - S_T, 0\},$$

where $K > 0$ is the *strike price* (or the *exercise price*) of the option. This option gives the right, but not the obligation, to *sell* the underlying asset at time T at the fixed price K. The payoff function for the option is

$$F^p(S) = \max\{K - S, 0\} = (K - S)^+.$$

11.3 No-Arbitrage Pricing: Main Result

The Pricing of Contingent Claims A central problem for the analysis of which the present model has been designed is as follows. Suppose we have a contingent claim paying $X(\omega)$ at time T—for example, a derivative security. What is the fair price of this contingent claim at time $t = 0$? In many cases, our model can give a clear and precise answer to this question. The principle underlying the asset pricing theory in this model is the principle of *pricing by no arbitrage*.

No Arbitrage Hypothesis Consider a trading strategy $\mathbf{H} = (\mathbf{h}_0, \ldots, \mathbf{h}_T)$, where $\mathbf{h}_0$ is the investor's portfolio at time 0, and $\mathbf{h}_t = \mathbf{h}_t(\omega^t)$ $(t = 1, 2, \ldots, T)$ is the portfolio held at time $t = 1, 2, \ldots, T$. Denote by $V_0^{\mathbf{H}}$ and $V_T^{\mathbf{H}}$ the values of the initial and the final portfolios for this strategy:

$$V_0^{\mathbf{H}} = \langle \mathbf{S}_0, \mathbf{h}_0 \rangle, \; V_T^{\mathbf{H}} = \langle \mathbf{S}_T, \mathbf{h}_T \rangle.$$

Note that $V_T^{\mathbf{H}} = V_T^{\mathbf{H}}(\omega)$ is random, while $V_0^{\mathbf{H}}$ is not.

We say that there is an *arbitrage opportunity* if there exists a self-financing trading strategy $\mathbf{H} = (\mathbf{h}_0, \ldots, \mathbf{h}_T)$ such that

$$V_0^{\mathbf{H}} \leq 0,$$

$$V_T^{\mathbf{H}}(\omega) \geq 0 \text{ for all } \omega \in \Omega \tag{11.3}$$

and

$$V_T^{\mathbf{H}}(\omega) > 0 \text{ for some } \omega \in \Omega. \tag{11.4}$$

It is said here that the self-financing strategy **H** makes it possible, starting from non-positive wealth at time 0, to get at time T non-negative wealth with probability one and strictly positive wealth with strictly positive probability: for at least one feasible market scenario ω. (Recall that Ω consists of those market scenarios ω for which $P(\omega) > 0$.)

We will impose the following fundamental assumption (the *no arbitrage hypothesis).*

(**NA**) The asset market under consideration does not allow for arbitrage opportunities.

Hedgeable Contingent Claims We say that a contingent claim $X(\omega)$ can be *replicated*, or *hedged*, if there exists a self-financing trading strategy $\mathbf{H} = (\mathbf{h}_0, \dots, \mathbf{h}_T)$ such that

$$\langle \mathbf{S}_T, \mathbf{h}_T \rangle = X(\omega) \text{ for all } \omega \in \Omega$$

(note that $\mathbf{S}_T$ and $\mathbf{h}_T$ depend on ω). This means that for all feasible market scenarios $\omega \in \Omega$ the strategy **H** yields exactly the same payoff at time T as the contingent claim $X(\omega)$.

No Arbitrage Pricing Principle It turns out that under the no arbitrage hypothesis, one can uniquely define a fair and economically justified price for every hedgeable contingent claim. This is done as follows. If X is a hedgeable contingent claim, then we can define its price at time 0 as the value $\langle \mathbf{S}_0, \mathbf{h}_0 \rangle$ of the initial portfolio $\mathbf{h}_0$ of any self-financing strategy **H** replicating X. As we shall see, under the no arbitrage hypothesis (**NA**), this value (remarkably!) does not depend on **H**.

Theorem 11.1 *Let assumption (NA) hold. If* $\mathbf{H} = (\mathbf{h}_0, \dots, \mathbf{h}_T)$ *and* $\mathbf{H}' = (\mathbf{h}'_0, \dots, \mathbf{h}'_T)$ *are two self-financing trading strategies replicating the same contingent claim* X*, then* $\langle \mathbf{S}_0, \mathbf{h}_0 \rangle = \langle \mathbf{S}_0, \mathbf{h}'_0 \rangle$.

The fact described in the above theorem is called *the law of one price*. As Theorem 11.1 shows, the law of one price follows from (**NA**). Thus, under hypothesis (**NA**), for each hedgeable contingent claim X, we can determine a well-defined level of initial wealth $w_0 = \langle \mathbf{S}_0, \mathbf{h}_0 \rangle$ needed at time 0 to hedge this contingent claim by means of some self-financing strategy $\mathbf{H} = (\mathbf{h}_0, \dots, \mathbf{h}_T)$. *According to the no arbitrage pricing principle,* w_0 *is the fair market price of the contingent claim* X *at time 0.*

Proof of Theorem 11.1 Suppose the contrary: there exist two self-financing trading strategies $\mathbf{H} = (\mathbf{h}_0, \ldots, \mathbf{h}_T)$ and $\mathbf{H}' = (\mathbf{h}'_0, \ldots, \mathbf{h}'_T)$ replicating the same contingent claim X, but

$$\langle \mathbf{S}_0, \mathbf{h}_0 \rangle \neq \langle \mathbf{S}_0, \mathbf{h}'_0 \rangle.$$

We may assume without loss of generality that $\langle \mathbf{S}_0, \mathbf{h}'_0 \rangle > \langle \mathbf{S}_0, \mathbf{h}_0 \rangle$, i.e., the difference $\Delta = \langle \mathbf{S}_0, \mathbf{h}'_0 \rangle - \langle \mathbf{S}_0, \mathbf{h}_0 \rangle$ is strictly positive. Consider the trading strategy $\mathbf{H}^\Delta = (\mathbf{h}^\Delta_0, \ldots, \mathbf{h}^\Delta_T)$, where

$$\mathbf{h}^\Delta_t = (\Delta, 0, 0, \ldots, 0)$$

for each t (the 0th coordinate of the vector $\mathbf{h}^\Delta_t$ is Δ, and all the other coordinates are equal to zero). Clearly the strategy $\mathbf{h}^\Delta_t$ is self-financing—because the portfolio $\mathbf{h}^\Delta_t$ does not change in time. Consequently, the strategy $\bar{\mathbf{H}} = (\bar{\mathbf{h}}_0, \ldots, \bar{\mathbf{h}}_T)$ defined by

$$\bar{\mathbf{h}}_t = \mathbf{h}_t - \mathbf{h}'_t + \mathbf{h}^\Delta_t$$

is self-financing. For this strategy, we have

$$V_0^{\bar{\mathbf{H}}} = \langle \mathbf{S}_0, \bar{\mathbf{h}}_0 \rangle = \langle \mathbf{S}_0, \mathbf{h}_0 \rangle - \langle \mathbf{S}_0, \mathbf{h}'_0 \rangle + \Delta = -\Delta + \Delta = 0 \tag{11.5}$$

and

$$\begin{aligned} V_T^{\bar{\mathbf{H}}} &= \langle \mathbf{S}_T, \bar{\mathbf{h}}_T \rangle = \langle \mathbf{S}_T, \mathbf{h}_T \rangle - \langle \mathbf{S}_T, \mathbf{h}'_T \rangle + \Delta(1+r)^T \\ &= X - X + \Delta(1+r)^T > 0 \text{ for all } \omega \in \Omega \end{aligned}$$

(we have used the fact that $S_t^0 = (1+r)^t$, $t = \overline{0, T}$). Thus the self-financing strategy $\bar{\mathbf{H}}$ provides an arbitrage opportunity: starting from zero wealth at time 0 it yields strictly positive wealth at time T for all $\omega \in \Omega$. This is a contradiction.

The proof is complete.

□

11.4 The No-Arbitrage Hypothesis and Net Present Value

Net Present Value In what follows we will need an equivalent version of hypothesis (**NA**) based on the notion of the *net present value*. To compare payoffs at time $t = T$ with those at time $t = 0$, we will use the discount factor $(1+r)^{-T}$. We define the *present value of the portfolio* $\mathbf{h}_T$ at time T as follows:

$$(1+r)^{-T} \langle \mathbf{S}_T, \mathbf{h}_T \rangle.$$

The *net present value* (NPV) of a trading strategy $\mathbf{H} = (\mathbf{h}_0, \ldots, \mathbf{h}_T)$ is defined as

$$\boxed{\text{present value of portfolio } \mathbf{h}_T} - \boxed{\text{value of portfolio } \mathbf{h}_0.}$$

In symbols,

$$V^{\mathbf{H}} = (1+r)^{-T} \langle \mathbf{S}_T, \mathbf{h}_T \rangle - \langle \mathbf{S}_0, \mathbf{h}_0 \rangle.$$

By using the notation $V_0^{\mathbf{H}} = \langle \mathbf{S}_0, \mathbf{h}_0 \rangle$ and $V_T^{\mathbf{H}} = \langle \mathbf{S}_T, \mathbf{h}_T \rangle$ for the values of the initial and the final portfolios of the strategy $\mathbf{H}$, we can rewrite the above formula as

$$V^{\mathbf{H}} = (1+r)^{-T} V_T^{\mathbf{H}} - V_0^{\mathbf{H}}.$$

Since $\mathbf{S}_T$ and $\mathbf{h}_T$ are functions of the market history $\omega = (a_1, \ldots, a_T)$, the net present value is a function of ω as well: $V^{\mathbf{H}} = V^{\mathbf{H}}(\omega)$. Let us consider the following version of the no arbitrage hypothesis:

(**NA1**) There is no self-financing trading strategy $\mathbf{H}$ for which the net present value $V^{\mathbf{H}}(\omega)$ is non-negative for all $\omega \in \Omega$ and strictly positive for some $\omega \in \Omega$.

Proposition 11.1 *Hypotheses (NA) and (NA1) are equivalent.*

Before proving Proposition 11.1, it is useful to establish the following auxiliary fact:

Proposition 11.2 *Let* $\mathbf{H}$ *be a self-financing strategy. Then there exists a self-financing strategy* $\tilde{\mathbf{H}}$ *such that*

$$V_0^{\tilde{\mathbf{H}}} = 0 \quad \textit{and} \quad V_T^{\tilde{\mathbf{H}}} = V^{\mathbf{H}}.$$

Proof of Proposition 11.2 Put $\Delta = \langle \mathbf{S}_0, \mathbf{h}_0 \rangle$ and consider the strategy $\mathbf{H}^{\Delta} = (\mathbf{h}_0^{\Delta}, \ldots, \mathbf{h}_T^{\Delta})$, where $\mathbf{h}_t^{\Delta} = (\Delta, 0, 0, \ldots, 0)$ for each t. The strategy $\mathbf{H}^{\Delta}$ is self-financing since $\mathbf{h}_t^{\Delta}$ does not change in time. Hence the strategy $\tilde{\mathbf{H}} = (\tilde{\mathbf{h}}_0, \ldots, \tilde{\mathbf{h}}_T)$, where

$$\tilde{\mathbf{h}}_t = (\mathbf{h}_t - \mathbf{h}_t^{\Delta})/(1+r)^T$$

is self-financing (verify!). For the strategy $\tilde{\mathbf{H}}$, we have

$$(1+r)^T V_0^{\tilde{\mathbf{H}}} = \langle \mathbf{S}_0, \mathbf{h}_0 \rangle - \langle \mathbf{S}_0, \mathbf{h}_0^{\Delta} \rangle = 0$$

and

$$V_T^{\tilde{\mathbf{H}}} = \frac{\langle \mathbf{S}_T, \mathbf{h}_T - \mathbf{h}_T^{\Delta} \rangle}{(1+r)^T} = \frac{\langle \mathbf{S}_T, \mathbf{h}_T \rangle}{(1+r)^T} - \Delta = V^{\mathbf{H}},$$

which completes the proof. □

Proof of Proposition 11.1 (NA) implies (NA1). Suppose (**NA1**) does not hold, i.e., there is a self-financing strategy $\mathbf{H} = (\mathbf{h}_0, \ldots, \mathbf{h}_T)$ such that $V^{\mathbf{H}}(\omega) \geq 0$ for all $\omega \in \Omega$ and $V^{\mathbf{H}}(\omega) > 0$ for some $\omega \in \Omega$. Consider the self-financing strategy $\tilde{\mathbf{H}}$ constructed in Proposition 11.2. Then $V_0^{\tilde{\mathbf{H}}} = 0$, $V_T^{\tilde{\mathbf{H}}}(\omega) = V^{\mathbf{H}}(\omega) \geq 0$ for all $\omega \in \Omega$ and $V_T^{\tilde{\mathbf{H}}}(\omega) = V^{\mathbf{H}}(\omega) > 0$ for some $\omega \in \Omega$. This contradicts (**NA**).

(NA1) implies (NA). Suppose (**NA**) does not hold, i.e., there is a self-financing $\mathbf{H}$ such that $V_0^{\mathbf{H}} \leq 0$, $V_T^{\mathbf{H}} \geq 0$ for all $\omega \in \Omega$ and $V_T^{\mathbf{H}} > 0$ for some $\omega \in \Omega$. But then $V^{\mathbf{H}} = (1+r)^{-T} V_T^{\mathbf{H}} - V_0^{\mathbf{H}} \geq 0$ for all $\omega \in \Omega$ and $V^{\mathbf{H}} > 0$ for some $\omega \in \Omega$. This contradicts (**NA1**).

The proof is complete. □

Risk-Neutral Pricing

12

12.1 Risk-Neutral Measures

The risk-neutral pricing principle plays a central role not only in connection with the problem of the pricing of derivative securities, but in Financial Economics as a whole. It has numerous applications. This principle may be regarded as a specification of the principle of pricing by no arbitrage as discussed in the previous chapter.

Risk-Neutral Probability Measure Recall that one of the main elements of the dynamic securities market model we deal with is the probability measure P on the set of all market histories $\omega = (a_1, a_2, \ldots, a_T)$. Recall that we denote by the letter E the expectation with respect to this probability measure P:

$$EX = \sum_{\omega \in \Omega} X(\omega)P(\omega),$$

where $X(\omega)$ is any function on Ω (random variable). The above sum can be taken over all feasible market histories $\omega \in \Omega$: for all the others, $P(\omega) = 0$. To formulate the risk-neutral pricing principle, we will need to consider other probability measures on Ω. Let Q be some other probability measure on Ω, i.e., a function $Q(\omega)$ of $\omega \in \Omega$ such that $Q(\omega) \geq 0$ and $\sum_{\omega \in \Omega} Q(\omega) = 1$. The expectation with respect to probability Q will be denoted by $E^Q X$:

$$E^Q X = \sum_{\omega \in \Omega} X(\omega)Q(\omega).$$

Definition of a Risk-Neutral Probability A probability measure Q on Ω is called *risk-neutral* if the expectation with respect to this measure of the net present value

I.V. Evstigneev et al., *Mathematical Financial Economics*, Springer Texts in Business and Economics, DOI 10.1007/978-3-319-16571-4_12

of any self-financing trading strategy is zero:

$$E^Q V^{\mathbf{H}} = 0. \tag{12.1}$$

Recall the definition of the net present value. For a trading strategy $\mathbf{H} = (\mathbf{h}_0, \ldots, \mathbf{h}_T)$, we have

$$V^{\mathbf{H}} = (1+r)^{-T}\langle \mathbf{S}_T, \mathbf{h}_T\rangle - \langle \mathbf{S}_0, \mathbf{h}_0\rangle = (1+r)^{-T} V_T^{\mathbf{H}} - V_0^{\mathbf{H}}. \tag{12.2}$$

Since $\mathbf{S}_0, \mathbf{h}_0$ and r are non-random, formula (12.1) can be written

$$(1+r)^{-T} E^Q V_T^{\mathbf{H}} - V_0^{\mathbf{H}} = 0,$$

or

$$V_0^{\mathbf{H}} = (1+r)^{-T} E^Q V_T^{\mathbf{H}}. \tag{12.3}$$

Note that the value $V_0^{\mathbf{H}}$ of the initial portfolio $\mathbf{h}_0$ of any self-financing trading strategy $\mathbf{H}$ is equal to the expectation—with respect to the risk-neutral measure Q—of the discounted value $V_T^{\mathbf{H}}/(1+r)^T$ of the final portfolio $\mathbf{h}_T$ of this strategy.

The property we have just formulated is an equivalent form of the definition of a risk-neutral probability measure: formula (12.3) is equivalent to formula (12.1).

Pricing by No Arbitrage and Risk-Neutral Pricing Recall the main idea of the principle of pricing by no arbitrage formulated in the previous chapter. This principle is applicable to establishing a fair price of any hedgeable contingent claim $X(\omega)$, in particular, a hedgeable derivative security. Recall that "hedgeable" means that there exists a self-financing trading strategy $\mathbf{H} = (\mathbf{h}_0, \ldots, \mathbf{h}_T)$ that hedges—or replicates—the contingent claim $X(\omega)$, i.e., yields the same payoff at time T as $X(\omega)$:

$$X(\omega) = V_T^{\mathbf{H}}(\omega) \text{ for all } \omega \in \Omega.$$

The principle of pricing by no arbitrage can be stated as follows:

Pricing by no arbitrage

The price of a contingent claim X is the value $V_0^{\mathbf{H}} = \langle \mathbf{S}_0, \mathbf{h}_0\rangle$ of the initial portfolio $\mathbf{h}_0$ of any self-financing strategy $\mathbf{H}$ replicating X.

There may be several self-financing strategies replicating $X(\omega)$, but, as we have shown in the previous chapter, if hypothesis (**NA**) holds, then the value $V_0^{\mathbf{H}} = \langle \mathbf{S}_0, \mathbf{h}_0\rangle$ of the initial portfolio is the same for all such strategies, and so the above price is, indeed, well-defined.

How to compute the value $V_0^{\mathbf{H}}$? The answer to this question can be given in terms of risk-neutral probabilities.

Recall that, for any risk-neutral probability Q and any self-financing trading strategy $\mathbf{H}$, we have

$$V_0^{\mathbf{H}} = (1+r)^{-T} E^Q V_T^{\mathbf{H}}.$$

Consequently, if $\mathbf{H}$ replicates the contingent claim X, i.e. $V_T^{\mathbf{H}} = X$, then

$$V_0^{\mathbf{H}} = (1+r)^{-T} E^Q X.$$

This leads to the following general pricing principle:

Risk-neutral pricing

The price of a contingent claim X is the discounted risk-neutral expected value $$(1+r)^{-T} E^Q X$$ of the contingent claim X.

Note that, although there may be several risk-neutral probability measures Q, the expected value $E^Q X$ is the same for all of them. Indeed, this expected value is equal to $(1+r)^T V_0^{\mathbf{H}}$ for any self-financing strategy $\mathbf{H}$ replicating X. We have seen that $V_0^{\mathbf{H}}$ does not depend on $\mathbf{H}$ and so the price $V_0^{\mathbf{H}} = (1+r)^{-T} E^Q X$ is determined uniquely by the given contingent claim X.

We emphasize that the pricing principles we deal with are applicable only to hedgeable contingent claims. In this connection the following notion is of importance.

Definition The market is called *complete* if any contingent claim is hedgeable.

In the specific models we will study (e.g., in those used for option pricing), the markets will be complete.

12.2 Fundamental Theorem of Asset Pricing

An Asset Pricing Theorem In connection with the risk-neutral pricing principle, the following questions arise:

- When can we guarantee that at least one risk-neutral probability measure exists?
- How to find this probability measure?
- How to verify that some given measure is indeed risk-neutral?

Fortunately, the first question has a very clear and general answer. This answer is contained in the celebrated result, first versions of which were obtained by Harrison, Kreps and Pliska in the 1970s, and which is referred to as the Fundamental Theorem of Asset Pricing.

Theorem 12.1 *If the market does not allow for arbitrage opportunities then a risk-neutral probability measure exists.*

Actually, in the present context all the theory is developed under the assumption that the no arbitrage hypothesis (**NA**) holds, and so the above result guarantees the existence of a risk neutral probability measure in those cases we are interested in.

Proof Suppose (**NA**), and hence (**NA1**), holds. Consider the following maximization problem:

$$\max_{\mathbf{H}} EV^{\mathbf{H}} \tag{12.4}$$

subject to

$$V^{\mathbf{H}}(\omega) \geq 0 \text{ for all } \omega \in \Omega. \tag{12.5}$$

Here the maximum is taken with respect to all self-financing trading strategies $\mathbf{H} = (\mathbf{h}_0, \mathbf{h}_1, \ldots, \mathbf{h}_T)$ satisfying constraints (12.5). We regard $\mathbf{H}$ as a vector whose coordinates are

$$h_t^i(\omega),\ i = 0, 1, \ldots, N,\ t = 0, 1, 2, \ldots, T,\ \omega \in \Omega.$$

For each $\omega \in \Omega$, the net present value $V^{\mathbf{H}}(\omega)$ is a linear function of $\mathbf{H}$, and so the constraints in (12.5) are linear. (The number of the constraints is equal to the number of points in Ω.) Also, $EV^{\mathbf{H}}$ is a linear function of $\mathbf{H}$, and so the maximization problem (12.4) and (12.5) is a linear programming problem.

Observe that the maximum value in the problem

$$\max_{\mathbf{H}} EV^{\mathbf{H}},$$

$$V^{\mathbf{H}}(\omega) \geq 0 \text{ for all } \omega \in \Omega,$$

is zero (it is attained at $\mathbf{H} = 0$). Suppose the contrary. Then there exists $\mathbf{H}$ such that $V^{\mathbf{H}}(\omega) \geq 0$ for all $\omega \in \Omega$ and $EV^{\mathbf{H}} > 0$. But then $V^{\mathbf{H}}(\omega) > 0$ for at least one $\omega \in \Omega$. This contradicts (**NA1**).

By using a general Lagrange multiplier rule (the Kuhn–Tucker theorem—see the Mathematical Appendix B), we construct Lagrange multipliers $\lambda(\omega) \geq 0$, $\omega \in \Omega$, such that

$$EV^{\mathbf{H}} + \sum_{\omega \in \Omega} \lambda(\omega) V^{\mathbf{H}}(\omega) \le 0$$

for all self-financing $\mathbf{H}$. We have zero in the right-hand side of this inequality since the maximum in (12.4) and (12.5) is zero.

The last inequality holds in fact as equality (replace $\mathbf{H}$ by $-\mathbf{H}$!), and so

$$\sum_{\omega \in \Omega} P(\omega) V^{\mathbf{H}}(\omega) + \sum_{\omega \in \Omega} \lambda(\omega) V^{\mathbf{H}}(\omega) = 0. \tag{12.6}$$

Put

$$M(\omega) = P(\omega) + \lambda(\omega),$$
$$M = \sum_{\omega \in \Omega} M(\omega),$$

and consider the probability measure Q defined for $\omega \in \Omega$ by

$$Q(\omega) = \frac{M(\omega)}{M}.$$

Thus, Q is the sought-for risk-neutral probability on Ω because $Q(\omega) > 0$,

$$\sum_{\omega \in \Omega} Q(\omega) = 1$$

and, by virtue of (12.6),

$$E^Q V^{\mathbf{H}} = \sum_{\omega \in \Omega} \frac{P(\omega) + \lambda(\omega)}{M} V^{\mathbf{H}}(\omega) = 0.$$

The proof is complete. □

Thus we have shown that, in the present context (where arbitrage opportunities are excluded), a risk-neutral probability measure always exists. Consequently, the first question has an affirmative answer.

12.3 Asset Pricing in Complete Markets

Complete Markets What can be said about the uniqueness of Q?

Theorem 12.2 *If the market is complete and does not allow for arbitrage opportunities, then a risk-neutral probability measure is unique.*

Proof Suppose the contrary: there are at least two different risk-neutral probabilities Q and Q'. Consider any market scenario $\omega^* \in \Omega$ and define the following contingent claim

$$X(\omega) = \begin{cases} 1 \text{ if } \omega = \omega^*, \\ 0 \text{ otherwise.} \end{cases}$$

Since the market is complete, this contingent claim is hedgeable. Consequently, the formulas

$$(1+r)^{-T} E^Q X \text{ and } (1+r)^{-T} E^{Q'} X$$

express the *uniquely defined* no arbitrage price of the contingent claim X. Thus

$$Q(\omega^*) = E^Q X = E^{Q'} X = Q'(\omega^*)$$

for each $\omega^* \in \Omega$, which means that Q and Q' coincide.

The contradiction proves the theorem. □

How to Verify Risk-Neutrality? Let us now turn to the second question: how can we construct a risk-neutral measure in an explicit form, i.e., compute it? To this end we have to find a convenient way of verifying the equality

$$E^Q V^{\mathbf{H}} = 0$$

for all self-financing strategies $\mathbf{H}$. (Recall that this equality is nothing but the definition of a risk-neutral probability Q.)

Having in mind the above goal, we will provide a formula establishing a relation between the net present value and *total discounted gain*.

Discounted Prices Let us introduce the following notation:

$$S_t = (S_t^1, \ldots, S_t^N),$$
$$s_t = (1+r)^{-t} S_t,$$
$$\mathbf{s}_t = (1+r)^{-t} \mathbf{S}_t.$$

Here S_t is the vector of prices of N risky assets, s_t is the vector of discounted prices of N risky assets, and $\mathbf{s}_t$ is the vector of discounted prices of $N+1$ assets (including the risk-free one):

$$\mathbf{s}_t = (1, s_t) = \left(1, \frac{S_t^1}{(1+r)^t}, \ldots, \frac{S_t^N}{(1+r)^t}\right).$$

The system of notation is as follows: lower-case letters denote discounted prices, and boldface is used for vectors of dimension $N + 1$ (to distinguish them from vectors of dimension N).

Discounted Gain Consider a trading strategy $\mathbf{H} = (\mathbf{h}_0, \ldots, \mathbf{h}_T)$. Recall that, for each t, $\mathbf{h}_t$ is a contingent portfolio

$$\mathbf{h}_t = \mathbf{h}_t(\omega^t) = \mathbf{h}_t(a_1, \ldots, a_t),$$

selected by the investor based on information about the market history $\omega^t = (a_1, \ldots, a_t)$ up to time t. Let us represent $\mathbf{h}_t$ in the form

$$\mathbf{h}_t = (h_t^0, h_t),$$

where h_t^0 is the risk-free component of $\mathbf{h}_t$ and $h_t = (h_t^1, \ldots, h_t^N)$ is the vector of risky assets contained in the portfolio $\mathbf{h}_t$. The scalar product

$$\langle h_t, s_{t+1} - s_t \rangle$$

is the *discounted gain* from the portfolio h_t in the time period $t, t+1$. The above expression is the difference between the value $\langle h_t, s_{t+1} \rangle$ of the portfolio h_t expressed in terms of the discounted prices $s_{t+1} = (1+r)^{-t-1} S_{t+1}$ and the value $\langle h_t, s_t \rangle$ of the portfolio h_t expressed in terms of the discounted prices $s_t = (1+r)^{-t} S_t$.

Net Present Value and Total Discounted Gain

Proposition 12.1 *For any self-financing trading strategy* $\mathbf{H}$*, the following formula holds:*

$$V^{\mathbf{H}} = \sum_{t=0}^{T-1} \langle h_t, s_{t+1} - s_t \rangle.$$

Thus the net present value of the self-financing trading strategy $\mathbf{H} = (\mathbf{h}_0, \ldots, \mathbf{h}_T)$ is equal to *the total discounted gain* obtained from the portfolios $h_0, \ldots, h_{T-1}$ of risky assets contained in the portfolios $\mathbf{h}_0, \ldots, \mathbf{h}_{T-1}$.

Proof of Proposition 12.1 The proof is an application of simple algebra:

$$\begin{aligned}
V^{\mathbf{H}} &= \frac{\langle \mathbf{S}_T, \mathbf{h}_T \rangle}{(1+r)^T} - \langle \mathbf{S}_0, \mathbf{h}_0 \rangle \\
&= \langle \mathbf{s}_T, \mathbf{h}_T \rangle - \langle \mathbf{s}_0, \mathbf{h}_0 \rangle \qquad (12.7) \\
&= \sum_{t=1}^{T} [\langle \mathbf{s}_t, \mathbf{h}_t \rangle - \langle \mathbf{s}_{t-1}, \mathbf{h}_{t-1} \rangle] \qquad (12.8) \\
&= \sum_{t=1}^{T} [\langle \mathbf{s}_t, \mathbf{h}_{t-1} \rangle - \langle \mathbf{s}_{t-1}, \mathbf{h}_{t-1} \rangle] \qquad (12.9) \\
&= \sum_{t=1}^{T} [s_t^0 h_{t-1}^0 - s_{t-1}^0 h_{t-1}^0] + \sum_{t=1}^{T} [\langle s_t, h_{t-1} \rangle - \langle s_{t-1}, h_{t-1} \rangle] \qquad (12.10) \\
&= \sum_{t=1}^{T} [\langle s_t, h_{t-1} \rangle - \langle s_{t-1}, h_{t-1} \rangle] = \sum_{t=0}^{T-1} \langle h_t, s_{t+1} - s_t \rangle, \qquad (12.11)
\end{aligned}$$

where equality (12.7) follows from the definition of the discounted prices $\mathbf{s}_t = (1+r)^{-t}\mathbf{S}_t$; (12.8) is obvious (cancel out terms with opposite signs!); (12.9) follows from the condition of self-financing; (12.10) is true since

$$\langle \mathbf{s}_t, \mathbf{h}_t \rangle = s_t^0 h_t^0 + \langle s_t, h_t \rangle;$$

(12.11) is valid because

$$s_t^0 = S_t^0 (1+r)^{-t} = 1.$$

Thus we have obtained that

$$V^{\mathbf{H}} = \sum_{t=0}^{T-1} \langle h_t, s_{t+1} - s_t \rangle \qquad (12.12)$$

for each self-financing strategy **H**, which terminates the proof. □

A Criterion for Risk-Neutrality The formula (12.12) leads to the following result providing sufficient (and in fact necessary) conditions for a probability measure Q to be risk-neutral.

Theorem 12.3 *If for every* $t = 0, 1, \dots, T-1$ *and for every contingent portfolio* $h_t = h_t(\omega^t)$ *of risky assets* $i = 1, 2, \dots, N$, *we have*

$$E^Q \langle h_t, s_{t+1} - s_t \rangle = 0, \qquad (12.13)$$

then the probability measure Q *is risk-neutral.*

Condition (12.13) means that the expectation E^Q of the discounted gain $\langle h_t, s_{t+1} - s_t \rangle$ is equal to zero for each contingent portfolio h_t. In the applications of the risk-neutral pricing principle, we will construct risk neutral probabilities Q by verifying condition (12.13) (which is much easier than dealing with the net present value directly).

Actually the assertion converse to Theorem 12.3 is valid as well, but it will not be needed in what follows.

Proof of Theorem 12.3 Straightforward: if (12.13) holds, then $E^Q V^{\mathbf{H}} = 0$ for each self-financing $\mathbf{H}$ [see (12.12)], and so Q is risk-neutral. □

Remark 12.2 Condition

$$E^Q \langle h_t, s_{t+1} - s_t \rangle = 0$$

holds for each $h_t(\omega^t)$ if and only if

$$E^Q(s_{t+1}|\omega^t) = s_t \text{ almost surely,} \tag{12.14}$$

where $E^Q(s_{t+1}|\omega^t)$ is the conditional expectation of s_{t+1} given ω^t. Property (12.14) means that the random process $s_0, s_1, \ldots, s_T$ is a *martingale* with respect to the probability measure Q. In this connection, risk-neutral probability measures are often called in the literature *martingale measures* (these terms pertain to the same class of measures).

13 The Cox–Ross–Rubinstein Binomial Model

13.1 The Structure of the Model

Model Description The *binomial model* was suggested by Cox, Ross and Rubinstein (1979) for the pricing of derivative securities. It is widely used for practical computations. In particular, it leads to very good approximations for known continuous-time option pricing formulas (such as the Black–Scholes one). Also, it demonstrates clearly the economic principles underlying such formulas.

The binomial model, which will be described now, is a special case of the general dynamic securities market model introduced in the previous chapters.

Assume that:

- The space A of states of the world consists of two elements, u and d ("up" and "down"), so that

$$A = \{u, d\}.$$

- There is one risk-free asset $i = 0$ (cash with constant interest rate $r \geq 0$) whose price at time t is $(1 + r)^t$.
- There is one risky security $i = 1$ whose price at time 0 is S_0 and price at time t is

$$S_t(a_1, a_2, \ldots, a_t),\ t = 1, 2, \ldots, T.$$

- The price process S_t has the following structure:

$$S_t = S_0 Z(a_1) Z(a_2) \ldots Z(a_t),$$

I.V. Evstigneev et al., *Mathematical Financial Economics*, Springer Texts in Business and Economics, DOI 10.1007/978-3-319-16571-4_13

where the values of the function $Z(a)$ on $A = \{u, d\}$ are two numbers

$$Z(d) < 1 < Z(u).$$

The Price Process The formula

$$S_t = S_0 Z(a_1) Z(a_2) \dots Z(a_t)$$

implies

$$S_t = S_{t-1} Z_t, \tag{13.1}$$

where

$$Z_t = Z(a_t).$$

Thus the price S_t at time t is equal to

$$S_t = \begin{cases} S_{t-1} Z(u) & \text{if } a_t = u, \\ S_{t-1} Z(d) & \text{if } a_t = d. \end{cases}$$

Since

$$Z(d) < 1 < Z(u),$$

the price can either "jump up", if the state of the world is u, or "jump down", if the state of the world is d.

In what follows, we will assume, additionally, that

$$Z(u) > 1 + r.$$

If the price jumps up, it increases more than the price of the risk-free security.

Note that, in this model, market scenarios (histories) $\omega \in \Omega$ are all possible sequences of length T of two symbols u and d—for example,

$$\omega = (u, u, d, d, d, u, d, u)$$

(here, $T = 8$).

In the present context, we will not specify the "real-world" probability measure P, that was introduced in the framework of the general dynamic securities market model (see Chap. 11). Its explicit form will not be needed for the pricing of derivative securities in the binomial model. What we need to know regarding this measure is only that all market scenarios ω have a strictly positive probability $P(\omega)$, so that the set *all* market scenarios coincides with the set Ω of *all feasible* ones.

13.2 Completeness of the Model

Hedging Contingent Claims in the Binomial Model We are going to demonstrate the use of the risk-neutral pricing principle in the present model. This principle assigns a unique and economically justified price to each hedgeable contingent claim. In this connection, the following question is important. Is the binomial model complete? In other words, can we hedge (replicate) *any* contingent claim? The answer is yes. The result is stated in the following theorem.

Theorem 13.1 *For any contingent claim $X(\omega)$ there exists a self-financing trading strategy $\mathbf{H} = (\mathbf{h}_0, \ldots, \mathbf{h}_T)$ such that $\langle \mathbf{S}_T, \mathbf{h}_T \rangle = X$ (for each ω).*

Proof Let us write $\mathbf{h}_t = (h_t^0, h_t)$, where h_t^0 is the risk-free component of the portfolio $\mathbf{h}_t$ and h_t is its risky component. We are going to construct the sought-for sequence of portfolios $\mathbf{h}_0, \ldots, \mathbf{h}_T$ from the end—starting from the last portfolio $\mathbf{h}_T$. Define

$$\mathbf{h}_T(\omega) = \left(\frac{X(\omega)}{(1+r)^T}, 0 \right).$$

Then $\langle \mathbf{S}_T, \mathbf{h}_T \rangle = X$.

Next, let us construct the portfolio $\mathbf{h}_{T-1} = \mathbf{h}_{T-1}(\omega^{T-1})$. The portfolio must satisfy the self-financing condition

$$\langle \mathbf{S}_T, \mathbf{h}_{T-1} \rangle = \langle \mathbf{S}_T, \mathbf{h}_T \rangle.$$

for each ω. This condition can be written

$$(1+r)^T h_{T-1}^0 + S_T h_{T-1} = X$$

where h_{T-1}^0 and h_{T-1} are functions of $\omega^{T-1} = (a_1, \ldots, a_{T-1})$, while S_T and X are functions of $\omega = (a_1, \ldots, a_T)$.

Recall that

$$S_T(a_1, \ldots, a_T) = S_{T-1}(a_1, \ldots, a_{T-1}) Z(a_T),$$

or

$$S_T(\omega^{T-1}, a_T) = S_{T-1}(\omega^{T-1}) Z(a_T)$$

for each partial history ω^{T-1} and each $a_T \in A = \{u, d\}$. Thus the self-financing condition

$$(1+r)^T h_{T-1}^0 + S_T h_{T-1} = X \tag{13.2}$$

can be written as a system of two linear equations

$$(1+r)^T h_{T-1}^0 + S_{T-1} Z(u) h_{T-1} = X(\omega^{T-1}, u),$$
$$(1+r)^T h_{T-1}^0 + S_{T-1} Z(d) h_{T-1} = X(\omega^{T-1}, d),$$

that must hold for each ω^{T-1}. Regarding h_{T-1}^0 and h_{T-1} as unknowns, we solve this system (ω^{T-1} being held fixed) and obtain the portfolio $\mathbf{h}_{T-1} = (h_{T-1}^0, h_{T-1})$ satisfying the self-financing condition

$$\langle \mathbf{S}_T, \mathbf{h}_{T-1} \rangle = \langle \mathbf{S}_T, \mathbf{h}_T \rangle \ (= X).$$

Then, quite analogously, we construct $\mathbf{h}_{T-2} = \mathbf{h}_{T-2}(\omega^{T-2})$ satisfying

$$\langle \mathbf{S}_{T-1}, \mathbf{h}_{T-2} \rangle = \langle \mathbf{S}_{T-1}, \mathbf{h}_{T-1} \rangle,$$

and so on. This procedure leads to the construction of a self-financing strategy $\mathbf{h}_0, \mathbf{h}_1, \ldots, \mathbf{h}_T$ with $\langle \mathbf{S}_T, \mathbf{h}_T \rangle = X$. □

Risk-Neutral Pricing in the Binomial Model According to the risk-neutral pricing principle, the fair price of any hedgeable contingent claim $X(\omega)$ is given by the formula

$$\text{price of } X = \frac{E^Q X}{(1+r)^T}, \text{ where } E^Q X = \sum_{\omega \in \Omega} Q(\omega) X(\omega),$$

Q being any risk-neutral probability measure. In the binomial model, every contingent claim is hedgeable (the model is complete), and so the above formula makes it possible to price all contingent claims. Thus it remains to construct a risk-neutral measure in the model at hand. (In fact it is unique, which follows from the market completeness.)

Recall that Ω is the set of all market histories

$$\omega = (a_1, \ldots, a_T), \ a_t \in A = \{u, d\}.$$

In order to specify a probability measure Q on Ω, we have to assign a non-negative number $Q(\omega)$ to each ω so that

$$\sum_{\omega \in \Omega} Q(\omega) = 1.$$

13.3 Constructing a Risk-Neutral Measure

An Explicit Formula for the Risk-Neutral Measure Q In the binomial model, we are given the following quantitative data:

- The risk-free interest rate $r \geq 0$.
- The numbers $Z(u)$ and $Z(d)$ specifying the possible price jumps—up and down, respectively.
- The initial value S_0 of the price of the risky asset.

Without loss of generality, we can assume—and we will assume from now on—that

$$Z(u) = u, \ Z(d) = d.$$

Thus the two elements of the set A are two numbers, u and d, characterizing the sizes of the price jumps.

Let $\omega = (a_1, a_2, \dots, a_T)$ be any market history. Define $Q(\omega)$ by the formula

$$Q(\omega) = q(a_1)q(a_2)\dots q(a_T), \tag{13.3}$$

where

$$q(a) = \begin{cases} p & \text{if } a = u, \\ 1 - p & \text{if } a = d. \end{cases} \tag{13.4}$$

The number p is defined by

$$p = \frac{1 + r - d}{u - d}. \tag{13.5}$$

Theorem 13.2 *The probability measure Q defined by (13.3), (13.4) and (13.5) is risk-neutral.*

Before proving the theorem, let us observe that the formulas (13.3)–(13.5) indeed determine a probability measure. First of all, the numbers $Q(\omega)$ are strictly positive. This is so because

$$p > 0 \text{ and } 1 - p > 0. \tag{13.6}$$

The former inequality is clear because we have assumed that the numbers $u = Z(u)$ and $d = Z(d)$ satisfy

$$d < 1, \ 1 + r < u. \tag{13.7}$$

The latter inequality in (13.6) can be written

$$1 + r - d < u - d,$$

which is equivalent to the latter inequality in (13.7).

It remains to check that the sum of $Q(\omega) = q(a_1)q(a_2)\dots q(a_T)$ over all $\omega = (a_1, \dots, a_T)$ is equal to one. This is so because

$$\sum_{\omega} Q(\omega) = \sum_{a_1=u,d} q(a_1) \sum_{a_2=u,d} q(a_2) \dots \sum_{a_T=u,d} q(a_T) = 1.$$

Proof of Theorem 13.2 Let us verify that the probability measure Q defined above is indeed risk-neutral. In the previous chapter, a method for this check was developed. According to this method, if we wish to verify that some probability measure Q is risk-neutral, we can proceed as follows:

- Consider the discounted prices

$$s_t = \frac{S_t}{(1+r)^t}, \; t = 0, 1, \dots, T,$$

  of the risky assets.
- For any moment of time $t = 0, 1, 2, \dots, T-1$, consider any contingent portfolio $h_t = h_t(\omega^t)$ of risky assets.
- If the expected discounted profit from this portfolio is zero,

$$E^Q \langle h_t, s_{t+1} - s_t \rangle = 0,$$

  then Q is risk-neutral.

Since in the present model there is only one risky security, we have to verify that

$$E^Q h_t (s_{t+1} - s_t) = 0,$$

for each function $h_t = h_t(\omega^t)$ (recall that $\omega^t = (a_1, \dots, a_t)$). To this end, we write

$$s_{t+1} - s_t = \frac{S_{t+1}}{(1+r)^{t+1}} - \frac{S_t}{(1+r)^t} = \frac{S_t}{(1+r)^t} \left(\frac{Z(a_{t+1})}{1+r} - 1 \right).$$

The last equality holds because $S_{t+1} = S_t Z(a_{t+1})$.

Thus we have to show that the random variable

$$h_t(s_{t+1} - s_t) = \frac{h_t S_t}{(1+r)^t} \left(\frac{Z(a_{t+1})}{1+r} - 1 \right) \tag{13.8}$$

has zero expectation with respect to Q. By virtue of the definition of Q, we have

$$Q(\omega) = q(a_1)q(a_2)\dots q(a_T) \text{ for each } \omega = (a_1, \dots, a_T).$$

Since $Q(\omega)$ is the product $q(a_1)q(a_2)\dots q(a_T)$, the random variables $a_1, a_2, \dots, a_T$ are *independent*. In formula (13.8), h_t and S_t are functions of $a_1, a_2, \dots, a_t$, while

$$W_{t+1} = \frac{Z(a_{t+1})}{1+r} - 1$$

is a function of a_{t+1}. Consequently, the random variables

$$h_t S_t \text{ and } W_{t+1}$$

are independent. To show that the expectation of (13.8) is zero, it is sufficient to show that

$$E^Q(h_t S_t W_{t+1}) = 0.$$

But the expectation of a product of independent random variables is equal to the product of their expectations:

$$E^Q(h_t S_t W_{t+1}) = E^Q(h_t S_t)\, E^Q W_{t+1}.$$

Thus it remains to verify that $E^Q W_{t+1} = 0$.

It is therefore left to show that

$$E^Q \left[\frac{Z(a_{t+1})}{1+r} - 1 \right] = 0$$

or, equivalently,

$$E^Q \frac{Z(a_{t+1})}{1+r} = 1.$$

If $\omega = (a_1, a_2, \dots, a_T)$ is distributed according to Q, the random variable a_{t+1} takes on the values u and d with probabilities p and $1 - p$, respectively. Therefore

$$E^Q \frac{Z(a_{t+1})}{1+r} = \frac{1}{1+r} [pZ(u) + (1-p)Z(d)] = \frac{pu + (1-p)d}{1+r}. \tag{13.9}$$

Recall the definition of p:

$$p = \frac{1+r-d}{u-d}.$$

By using this definition, we write

$$pu + (1-p)d = d + p(u-d) = d + 1 + r - d = 1 + r,$$

which, combined with (13.9), yields the desired equality

$$E^Q \frac{Z(a_{t+1})}{1+r} = 1.$$

The proof is complete. □

Remark 13.1 If $\omega = (a_1, \dots, a_T)$ is distributed according to the probability Q just constructed, then the states of the world $a_1, \dots, a_T$ are *independent and identically distributed (i.i.d.)*. This follows from formula (13.3) defining Q.

The Pricing of Derivative Securities The results obtained for the binomial model make it possible to compute numerically the prices of various derivative securities. Recall that each derivative security is characterized by its payoff function (its value at time T)

$$X = F(S_0, S_1, \dots, S_T),$$

depending on the prices $S_0, S_1, \dots, S_T$ of the underlying asset. Our goal is to compute the fair price of the derivative security at time 0. To this end we regard $X = X(\omega)$ as a contingent claim (X depends on the market history $\omega \in \Omega$, since the prices $S_0, S_1, \dots, S_T$ depend on ω), and use the formula

$$\text{price} = \frac{E^Q X}{(1+r)^T},$$

where Q is the risk-neutral probability measure on Ω.

13.4 Examples

Examples of Derivative Securities A few examples of different derivative securities are provided below.

European Derivative Securities The payoff function of a derivative security of this kind depends only on the price S_T of the underlying asset at time T. This is so, for example, for already mentioned *European call and put options*, for which the payoff functions $F^c(S_T)$ and $F^p(S_T)$ are given by

$$F^c(S_T) = (S_T - K)^+, \; F^p(S_T) = (K - S_T)^+,$$

where $K > 0$ is the *strike price* of the option. The European *call* (*put*) option gives the right, but does not impose the obligation, to *buy* (*sell*) the underlying asset at time T at the fixed price K.

Asian Options Examples of derivative securities for which the payoff function depends on the whole sequence of prices $S_0, \dots, S_T$ are *Asian options*, involving the *average* price of underlying asset:

$$\bar{S}_T = (S_0 + \ldots + S_T)/(T + 1).$$

In particular, the *average strike call and put options* have the payoff functions

$$F_{as}^c = (S_T - \bar{S}_T)^+, \ F_{as}^p = (\bar{S}_T - S_T)^+,$$

where the role of the strike price is played by $\bar{S}_T$.

Lookback Options *Lookback call* allows one to *buy* the risky asset at the *smallest* of the prices $S_0, \dots, S_T$. The (non-negative) payoff function is

$$F_l^c = S_T - \min_{0 \le t \le T} S_t.$$

Analogously, a *lookback put* allows one to *sell* the risky asset at the *greatest* of the prices $S_0, \dots, S_T$, i.e., the (non-negative) payoff is

$$F_l^p = \max_{0 \le t \le T} S_t - S_T.$$

Numerical examples related to the pricing of these and other derivative securities will be considered in Chap. 16.

Two-Period Binomial Model Suppose, for example, it is required to compute the price at time 0 of a derivative security, characterized by its payoff function $F(S_T)$, in a two-period binomial model ($T = 2$). In the model, we are given:

- The risk-free interest rate $r \ge 0$.
- The numbers u and d specifying the possible price jumps—up and down, respectively.
- The price S_0 of the risky asset at time 0.

The algorithm for solving the problem is as follows. We first compute the number

$$p = \frac{1 + r - d}{u - d}. \tag{13.10}$$

This number determines the risk-neutral measure $Q(\omega)$ by the formula

$$Q(\omega) = q(a_1)q(a_2), \ \omega = (a_1, a_2),$$

where

$$q(a) = \begin{cases} p & \text{if } a = u, \\ 1 - p & \text{if } a = d. \end{cases}$$

In this model there are four market scenarios

$$(u, u), \ (u, d), \ (d, u), \ (d, d),$$

occurring with probabilities

$$p^2, \ p(1-p), \ (1-p)p, \ (1-p)^2.$$

The table below indicates possible market histories $\omega \in \Omega$ (the set Ω consists of four points), the probabilities $Q(\omega)$ and corresponding value of the price $S_T = S_T(\omega)$ (Table 13.1).

The risk-neutral expected payoff of the contingent claim $X(\omega) = F(S_T)$ is as follows:

$$E^Q X = p^2 F(S_0 u^2) + 2p(1-p)F(S_0 ud) + (1-p)^2 F(S_0 d^2).$$

The price of the derivative security is given by:

$$\text{price} = \frac{E^Q X}{(1+r)^2} = \frac{p^2 F(S_0 u^2) + 2p(1-p)F(S_0 ud) + (1-p)^2 F(S_0 d^2)}{(1+r)^2}.$$

***T*-Period Binomial Model** To conclude this chapter consider a T-period binomial model with parameters r, u, d and S_0, where $T > 2$. The price $S_T(\omega)$ can take on any of the $T + 1$ values

$$S_0 u^j d^{T-j}, \ j = 0, \dots, T.$$

Table 13.1 Market histories, their probabilities and prices

ω	$Q(\omega)$	$S_T(\omega)$
(u, u)	p^2	$S_0 u^2$
(u, d)	$p(1-p)$	$S_0 ud$
(d, u)	$(1-p)p$	$S_0 ud$
(d, d)	$(1-p)^2$	$S_0 d^2$

The value $S_0 u^j d^{T-j}$ corresponds to those histories $\omega = (a_1, \ldots, a_T)$ in which exactly j symbols a_t are equal to u. There are

$$C_j^T = \frac{T!}{j!(T-j)!}$$

($j! = 1 \cdot 2 \cdot 3 \cdot \ldots \cdot j$) such histories ω, and each of such ω occurs with probability $p^j(1-p)^{T-j}$ where p is defined in (13.10).

The price of a derivative security with payoff function $F(S_T)$ is therefore given by

$$E^Q F(S_T) = \sum_{j=0}^{T} p^j (1-p)^{T-j} C_j^T F(S_0 u^j d^{T-j})$$

which equates to

$$\text{price} = \frac{1}{(1+r)^T} \sum_{j=0}^{T} p^j (1-p)^{T-j} C_j^T F(S_0 u^j d^{T-j}).$$

American Derivative Securities 14

14.1 The Notion of an American Derivative Security

The Basics Consider a T-period dynamic securities market model with two assets—risk-free and risky. Let $r > 0$ be the risk-free interest rate and let S_t ($t = 0, 1, \ldots, T$) be the price of the risky asset at time t. It is supposed that, for $t \geq 1$, the price S_t depends on the sequence ω^t of the states of the world $a_1, a_2, \ldots, a_t$. We will assume that the market under consideration is complete and does not allow for arbitrage opportunities.

In the previous chapters, a theory was developed providing foundations for the pricing of contingent claims (derivative securities)

$$X(\omega),\ \omega = (a_1, \ldots, a_T),$$

the payoff from which is obtained at time T—at the end of the time interval under consideration. Examples of derivative securities of this kind are European call and put options:

$$X^c = \max\{S_T - K, 0\} = (S_T - K)^+,\ X^p = \max\{K - S_T, 0\} = (K - S_T)^+,$$

where $K > 0$ is the *strike price* (or the *exercise price*) of the option. Recall that an option of the first kind gives the right to *buy* the underlying asset at time T at the fixed price K. An option of the second kind gives the right to *sell* the underlying asset at time T at the price K.

American Options and Other American Derivative Securities In real financial markets, an important role is played by *American options*, which allow one to buy or sell the underlying asset at the given price at any moment of time τ between 0 and

I.V. Evstigneev et al., *Mathematical Financial Economics*, Springer Texts in Business and Economics, DOI 10.1007/978-3-319-16571-4_14

T.The payoffs obtained by an owner of such options are as follows:

$$\max\{S_\tau - K, 0\} = (S_\tau - K)^+, \ \max\{K - S_\tau, 0\} = (K - S_\tau)^+.$$

These payoffs, depend of course on the moment of time τ at which the owner of the option decides to exercise it.

American options are examples of *American derivative securities*: derivative securities that can be exercised at any moment of time $0 \le \tau \le T$. An American derivative security is defined in terms of its payoff function $F(S)$. If we exercise it at time τ, we get $F(S_\tau)$.

To provide an algorithm for the pricing of a conventional derivative security (whose payoff is obtained at time T) it is sufficient to know its payoff function. This is not the case with an American derivative security. For a complete description of it (needed for computing its price), we have to specify not only its payoff function but also the *exercise strategy* which indicates the moment of time τ when the owner of the derivative security is going to exercise it.

Exercise Strategies An *exercise strategy* of an American derivative security is a function $\tau = \tau(S_0, S_1, \dots, S_T)$ of the sequence $(S_0, S_1, \dots, S_T)$ of the prices of the underlying security indicating the moment of time $0 \le \tau \le T$ when the derivative security should be exercised. It is natural to consider only those exercise strategies τ that are *non-anticipative* ("independent of the future") in the following sense:

(*) If $\tau(S_0, S_1, \dots, S_T) = t$ for some sequence of prices $S_0, S_1, \dots, S_T$, and if $S'_0, S'_1, \dots, S'_T$ is another sequence of prices such that

$$S_k = S'_k \text{ for all } k \le t,$$

then $\tau(S'_0, S'_1, \dots, S'_T) = t$.

Property (*) will be *included* into the definition of an exercise strategy.[1]

It is assumed that the owner of the derivative security observes the prices $S_0, S_1, S_2, \dots, S_T$ and decides to "stop" holding the contract at the moment of time τ prescribed by the exercise strategy $\tau(S_0, S_1, \dots, S_T)$. At this moment, the payoff $F(S_\tau)$ is received. Non-anticipativity expresses the idea that the decision to stop or not to stop at some moment of time t depends only on the observation of the past and the present of the price process and does not depend on its future, $S_{t+1}, S_{t+2}, \dots, S_T$.

[1]The notion of an non-anticipative exercise strategy is closely related to the notions of a "stopping time" and an "optional time" used in the literature.

14.2 Risk-Neutral Pricing of American Derivative Securities

The Risk-Neutral Pricing Principle for American Derivative Securities This principle is as follows.

> The risk-neutral price of an American derivative security with payoff function $F(S)$ and exercise strategy τ is given by
> $$E^Q \frac{1}{(1+r)^\tau} F(S_\tau),$$
> where Q is the risk-neutral probability.

Recall that we have assumed that the market under consideration is complete and does not have arbitrage opportunities. Consequently, the risk-neutral probability measure Q on the space Ω of the market scenarios $\omega = (a_1, a_2, \ldots, a_T)$ exists and is unique.

Justification of the Risk-Neutral Pricing Principle for American Derivative Securities To justify the above pricing principle we use the following consideration. The payoff $F(S_\tau)$ received at some time $0 \le \tau \le T$ is equivalent to the payoff $(1+r)^{T-\tau} F(S_\tau)$ received at time T. Indeed, in the model we deal with, there is a risk-free asset (cash) with constant interest rate $r \ge 0$. The American derivative security costs $F(S_\tau)$ at time τ when it is exercised. Its owner can sell it at time τ and deposit the amount $F(S_\tau)$ with the bank, which will yield (taking into account the interest rate r) the amount

$$X(\omega) = (1+r)^{T-\tau} F(S_\tau)$$

at time T.

Thus the American derivative security with payoff function $F(S_\tau)$ exercised at time τ is equivalent to the "conventional" derivative security

$$X(\omega) = (1+r)^{T-\tau} F(S_\tau),$$

the payoff from which is obtained at time T.

Note that the moment of time $\tau = \tau(S_0, \ldots, S_T)$ is a function of $\omega = (a_1, \ldots, a_T)$, since, for each t, the price $S_t = S_t(a_1, \ldots, a_t)$ is a function of the market history $\omega^t = (a_1, \ldots, a_t)$ up to time t. Hence both τ and S_τ in the above formula are functions of ω.

We have assumed that the market under consideration is complete and does not allow for arbitrageopportunities. Therefore the risk-neutral price of the conventional derivative security

$$X(\omega) = (1+r)^{T-\tau} F(S_\tau)$$

is given by the discounted expected value of $X(\omega)$ with respect to the risk-neutral probability Q:

$$\frac{1}{(1+r)^T} E^Q X(\omega) = \frac{1}{(1+r)^T} E^Q (1+r)^{T-\tau} F(S_\tau) = E^Q \frac{1}{(1+r)^\tau} F(S_\tau),$$

which leads to the risk-neutral pricing principle as stated above.

The Upper Price of an American Derivative Security The maximum of all risk-neutral prices over all exercise strategies τ

$$\max_\tau E^Q \frac{1}{(1+r)^\tau} F(S_\tau)$$

is called the *upper price* (or the *seller's price*) of the American derivative security. Generally, the seller of the derivative security does not know which particular exercise strategy τ the buyer is going to use. Therefore the price set by the seller is the maximum of all the risk-neutral prices over all exercise strategies τ. (Recall that the risk-neutral price of a contingent claim is equal to the level of initial wealth needed to replicate this contingent claim.)

American Call Options and European Call Options Suppose the derivative security under consideration is the *American call option* with strike price K and payoff

$$F(S) = \max\{S-K, 0\} = (S-K)^+.$$

It can be shown that, for any exercise strategy τ,

$$E^Q \frac{F(S_\tau)}{(1+r)^\tau} \leq E^Q \frac{F(S_T)}{(1+r)^T},$$

as long as $F(S) = (S-K)^+$. Thus—as $\tau = T$ is an exercise strategy leading to the payoff $F(S_T)$—*the upper risk-neutral price of the American call option is equal to the risk-neutral price of the European call option with the same strike price*. For a proof of this statement see Chap. 16, Question 16.2.

The analogous result for American put options is not valid. One cannot reduce the problem of their pricing to the analogous problem for European options. An algorithm suitable for this purpose will be developed in the remainder of this chapter.

The Conditional Value of an American Derivative Security Let us describe a method for the pricing of American derivative securities in a binomial model. Consider a T-period binomial model with parameters u, d, r and S_0. The price S_t in this model can take on $t + 1$ values:

$$S_{t,j} = S_0 u^j d^{t-j}, \; j = 0, 1, \ldots, t.$$

For each $t = 0, 1, \ldots, T$, and each $j = 0, 1, 2, \ldots, t$, define

$$V_t(j) = \max_{t \leq \tau \leq T} E^Q \left[\frac{F(S_\tau)}{(1+r)^\tau} \mid S_t = S_{t,j} \right].$$

Here, we consider the conditional expectation (with respect to the risk-neutral measure Q) of the discounted payoff

$$\frac{F(S_\tau)}{(1+r)^\tau}$$

assuming that the price S_t of the risky asset at time t is $S_{t,j}$. We take the maximum of such conditional expectations over all exercise strategies

$$\tau = \tau(S_t, S_{t+1}, \ldots, S_T)$$

satisfying $t \leq \tau \leq T$ and depending only on the prices $S_t, S_{t+1}, \ldots, S_T$. The number $V_t(j)$ is the *maximum expected discounted payoff that can be attained when the derivative security is exercised at time* $t \leq \tau \leq T$ *and when* $S_t = S_{t,j}$.

We will call $V_t(j)$ *the conditional value* (given the price $S_{t,j}$ at time t) of the derivative security under consideration.

Proposition 14.1 *The conditional value* $V_0(0)$ *coincides with the upper price of the American derivative security.*

Proof If $t = 0$, there is only one price of the form $S_{t,j} = S_0 u^j d^{t-j}$ ($j = 0, 1, \ldots, t$). This is the given initial price of the risky asset,

$$S_{t,j} = S_{0,0} = S_0.$$

Consequently, if $t = 0$, the conditional expectation involved in the definition of $V_t(j)$,

$$V_t(j) = \max_{t \leq \tau \leq T} E^Q \left[\frac{F(S_\tau)}{(1+r)^\tau} \mid S_t = S_{t,j} \right],$$

coincides with the unconditional expectation

$$E^Q\left[\frac{F(S_\tau)}{(1+r)^\tau}\right].$$

The maximum of such expectations with respect to all exercise strategies $0 \leq \tau(S_0, \ldots, S_T) \leq T$ is, by definition, the upper price of the American derivative security under consideration. □

14.3 The Pricing Algorithm

Computing the Seller's Price by Backward Induction Define

$$p = \frac{1+r-d}{u-d}, \quad F_t(S) = \frac{F(S)}{(1+r)^t}.$$

Here, p denotes the probability with which the price of the risky asset in the binomial model goes up, as long as the price process is governed by the risk-neutral probability measure Q. We denote by $F_t(S)$ the payoff function $F(S)$ discounted by the factor $(1+r)^{-t}$.

Proposition 14.2 *For the conditional value $V_T(j)$ at the last moment of time T, we have:*

$$V_T(j) = F_T(S_{T,j}), \ j = 0, 1, \ldots, T. \tag{14.1}$$

The conditional value $V_t(j)$ can be expressed through the conditional values $V_{t+1}(j+1)$, $V_{t+1}(j)$ and the payoff $F_t(S_{t,j})$ as follows:

$$V_t(j) = \max\{F_t(S_{t,j}), \ pV_{t+1}(j+1) + (1-p)V_{t+1}(j)\}. \tag{14.2}$$

This proposition suggests an *algorithm for computing $V_0(0)$—the upper price of the American derivative security*. First compute the numbers

$$V_T(j) = F_T(S_{T,j}), \ j = 0, 1, \ldots, T.$$

Then, by using formula (14.2), compute step by step

$$V_{T-1}(j), \ j = 0, 1, \ldots, T-1,$$

$$V_{T-2}(j), \ j = 0, 1, \ldots, T-2,$$

and so on. At the final step, the number $V_0(0)$ is obtained. This is the upper price of the American derivative security.

Proof of Proposition 14.2

1st step. Let us first prove formula (14.1). Recall the definition of the conditional value $V_t(j)$:

$$V_t(j) = \max_{t \le \tau \le T} E^Q \left[\frac{F(S_\tau)}{(1+r)^\tau} \mid S_t = S_{t,j} \right], \tag{14.3}$$

where the maximum is taken over all exercise strategies $\tau = \tau(S_t, S_{t+1}, \dots, S_T)$ satisfying $t \le \tau \le T$. The formula (14.1) in Proposition 14.2, which we have to prove, is as follows:

$$V_T(j) = F_T(S_{T,j}), \ j = 0, 1, \dots, T.$$

This formula is immediate from (14.3): if $t = T$, then the only exercise strategy τ satisfying $t \le \tau \le T$ is $\tau = T$, and so the maximum in (14.3) reduces to

$$\frac{F(S_{T,j})}{(1+r)^T} = F_T(S_{T,j}),$$

which proves (14.1).

2nd step. To prove the second formula in Proposition 14.2 let us observe the following: any non-anticipative strategy $t \le \tau(S_t, \dots, S_T) \le T$ either prescribes to exercise the derivative security at time t for any sequence of prices $S_t, \dots, S_T$ with $S_t = S_{t,j}$ or prescribes to exercise the derivative security *later* than t for any such sequence of prices. This follows from the non-anticipativity of $\tau(S_t, \dots, S_T)$—recall that we have included this property into the definition of all exercise strategies!

Indeed, suppose

$$\tau(S_t, \dots, S_T) = t$$

for some sequence of prices $S_t, \dots, S_T$ with $S_t = S_{t,j}$. Consider another sequence of prices $S'_t, \dots, S'_T$ with $S'_t = S_{t,j}$. Then $S_k = S'_k$ for all $k \le t$, which means here $S_k = S'_k$ for $k = t$. By the definition of non-anticipativity of a strategy, we have $\tau(S'_t, \dots, S'_T) = t$.

Suppose now that

$$\tau(S_t, \dots, S_T) > t$$

for some sequence of prices $S_t, \dots, S_T$ with $S_t = S_{t,j}$. Consider another sequence of prices $S'_t, \dots, S'_T$ with $S'_t = S_{t,j}$. Assume $\tau(S'_t, \dots, S'_T) = t$. But in this case $\tau(S_t, \ldots, S_T) = t$, as we have shown above. A contradiction. Thus $\tau(S'_t, \ldots, S'_T) > t$.

3rd step. Let us prove formula (14.2):

$$V_t(j) = \max\{F_t(S_{t,j}),\ pV_{t+1}(j+1) + (1-p)V_{t+1}(j)\}.$$

Recall that

$$V_t(j) = \max_{t \le \tau \le T} E^Q\left[\frac{F(S_\tau)}{(1+r)^\tau} \mid S_t = S_{t,j}\right],$$

We have shown that for any strategy $t \le \tau(S_t, \ldots, S_T) \le T$ either $\tau = t$ for any sequence of prices $S_t, \ldots, S_T$ with $S_t = S_{t,j}$ or $\tau \ge t+1$ for any such sequence of prices.

If $\tau = t$, the discounted payoff is $F_t(S_{t,j})$.

Suppose $\tau \ge t+1$. To analyze this case we note that the price S_{t+1} at time $t+1$ can be either

$$S_{t,j}u = (S_0 u^j d^{t-j})u = S_0 u^{j+1} d^{t-j} = S_0 u^{j+1} d^{(t+1)-(j+1)} = S_{t+1,j+1}$$

with probability p or

$$S_{t,j}d = (S_0 u^j d^{t-j})d = S_0 u^j d^{t-j+1} = S_{t+1,j}$$

with probability $1-p$.

If $S_{t+1} = S_{t+1,j+1}$, then the maximum expected discounted payoff we can obtain by exercising the derivative security at time $t+1 \le \tau \le T$ is $V_{t+1}(j+1)$, and if $S_{t+1} = S_{t+1,j}$, then the maximum expected discounted payoff is $V_{t+1}(j)$. Since $S_{t+1} = S_{t+1,j+1}$ with probability p and $S_{t+1} = S_{t+1,j}$ with probability $1-p$, the maximum expected discounted payoff is $pV_{t+1}(j+1)+(1-p)V_{t+1}(j)$. By taking the maximum of this expression and $F_t(S_{t,j})$, we arrive at the formula (14.2) asserted in the proposition.

□

From Binomial Model to Black–Scholes Formula

15

15.1 Drift and Volatility

Drift The purpose of this chapter is to derive the Black–Scholes formula—one of the highlights of Mathematical Finance. To this end we will use, as an auxiliary tool, the binomial model considered in the previous chapter. We begin with introducing the notions of the drift and the volatility of a price process which will be used in the further analysis.

Let $S(t)$ denote the price of some asset at time $t \geq 0$ (here, the time t can be either continuous or discrete). It is assumed that this price is random, and so the random function $S(t)$ is a *stochastic (random) process*. There are two important characteristics of the price process $S(t)$ that can be estimated statistically and then used for financial decisions—the drift and the volatility.

Definition The price process $S(t)$ is said to have a *(constant) drift* μ if

$$E \ln \frac{S(t+m)}{S(t)} = \mu m \quad \text{for all } t \geq 0 \text{ and } m \geq 0.$$

This means that the expected logarithm of the percentage change

$$\frac{S(t+m)}{S(t)}$$

of the price over the time interval $t,\ t+m$ is *proportional* to the length m of the time interval, the coefficient of proportionality being equal to μ. If $\mu > 0$, the price grows on average; if $\mu < 0$, it tends to decrease.

I.V. Evstigneev et al., *Mathematical Financial Economics*, Springer Texts in Business and Economics, DOI 10.1007/978-3-319-16571-4_15

Volatility

Definition We say that the price process $S(t)$ has a *(constant) volatility* σ if

$$Var\left[\ln \frac{S(t+m)}{S(t)}\right] = m\,\sigma^2 \quad \text{for all } t \geq 0 \text{ and } m \geq 0.$$

This formula means that the variance of the percentage change

$$\frac{S(t+m)}{S(t)}$$

of the price (in the logarithmic scale) over a time interval of length m is *proportional* to m, the coefficient of proportionality being equal to σ^2. The greater the volatility, the more "variable" is the price $S(t)$, and the more risky is the investment in the asset.

15.2 Modelling the Price Process

Empirical Drift and Volatility Suppose we have estimated—by using statistical methods—the drift and the volatility of a stochastic process $S(t)$ describing the price of a security traded in the market. Based on this information, we have to make financial decisions and provide quantitative estimates. For example, it is required to compute the price of a derivative of the security with price $S(t)$, say, a European call option with given strike price. To solve this problem, we need a *model* for the random process $S(t)$—we have to accept some reasonable hypotheses about the laws governing the time evolution of the price $S(t)$.

Geometric Random Walk A very common hypothesis is that the price $S(t)$ observed at some discrete moments of time

$$t_0 = 0,\ t_1 = \Delta,\ t_2 = 2\Delta,\ t_3 = 3\Delta,\ \ldots$$

in equal time intervals of length $\Delta > 0$, follows a *geometric random walk*. This means that the random price $S(t_k)$ can be represented in the form

$$S(t_k) = S_0 Z_1 Z_2 \ldots Z_k,$$

where $Z_1, Z_2, \ldots, Z_k$ are independent and identically distributed (i.i.d.) random variables. Equivalently, this means that the returns on the asset

$$\frac{S(t_k) - S(t_{k-1})}{S(t_{k-1})} = Z_k - 1$$

over the intervals t_{k-1}, t_k are i.i.d. random variables. Such price processes can be generated by using the binomial model. We will show how this can be done with the view to practical computations.

Continuous Compounding Suppose that in the financial market we are modelling, the nominal interest rate per unit of time (1 year) is r. Further, the interest is compounded *continuously*. Then one Euro deposited in the bank account at the beginning of a time period of length T will yield e^{rT} at the end of the period. This conclusion is justified by the following argument. Suppose the time interval $[0, T]$ is divided into n equal parts and the interest is compounded at each of the moments of time

$$t_0 = 0,\ t_1 = \frac{T}{n},\ t_2 = \frac{2T}{n},\ \ldots,\ t_{n-1} = \frac{(n-1)T}{n},\ t_n = T.$$

Then if we deposit one Euro at time $t_0 = 0$, we get

$$1 + \frac{rT}{n} \text{ at time } t_1,$$

$$\left(1 + \frac{rT}{n}\right)^2 \text{ at time } t_2,$$

and so on. Finally, at time $t_n = T$, we obtain $(1 + rT/n)^n$. We have

$$\left(1 + \frac{rT}{n}\right)^n \approx e^{rT} \text{ for large } n,$$

which justifies the formula e^{rT} for the continuously compounded interest.[1]

15.3 Binomial Approximation of the Price Process

Discretization of the Price Process Suppose we wish to analyze a price process $S(t)$ over some given time interval $[0, T]$. Fix n for the moment and divide the segment $[0, T]$ into n equal segments by the points

$$t_0 = 0,\ t_1 = \Delta,\ t_2 = 2\Delta,\ \ldots,\ t_{n-1} = (n-1)\Delta,\ t_n = n\Delta = T,$$

[1] We use here (for $x = rT$) the fact that $(1 + x/n)^n \to e^x$ as $n \to \infty$.

where

$$\Delta = T/n.$$

Assume $S(t)$ can change only at times $t = t_1, t_2, \ldots, t_n$, and the change is either u with probability $\pi > 0$ or d with probability $1 - \pi$, where $d < 1 < u$. Formally, consider the binomial model in which:

- The space A of states of the world consists of two elements—the numbers u and d.
- The price of the single risky security $i = 1$ at time 0 is $S(0) = S_0$ and its price $S(t_k)$ at time t_k—which we will briefly denote by S_k—is $S_k(a_1, a_2, \ldots, a_k)$, $k = 1, 2, \ldots, n$.
- The price process S_k has the following structure:

$$S_k = S_0 Z(a_1) Z(a_2) \ldots Z(a_k),$$

  where the values of the function $Z(a)$ on $A = \{u, d\}$ are $Z(u) = u$ and $Z(d) = d$.
- There is one risk-free asset $i = 0$—cash with constant interest rate $r_n = r\Delta$.
- The probability measure $P(\omega)$, $\omega = (a_1, a_2, \ldots, a_n)$, is given by $P(\omega) = \pi(a_1) \ldots \pi(a_n)$, where $\pi(u) = \pi$ and $\pi(d) = 1 - \pi$ (this means that $a_1, \ldots, a_n$ are independent, $P\{a_k = u\} = \pi$ and $P\{a_k = d\} = 1 - \pi$).

Selecting Parameters of the Binomial Approximation Suppose our statistical data show that the previous behavior of the process $S(t)$ has been characterized by a constant drift μ and a constant volatility σ, and we do not expect any changes in σ and μ during the time interval $[0, T]$. Further, assume that the nominal interest rate per unit time is r, and the interest is compounded continuously. We would like to define the parameters of the binomial model so that, in the limit as $n \to \infty$ (i.e., when the partition of the time interval $[0, T]$ becomes finer and finer), the model will match the given data.

We define S_0 as the observed price $S(0)$ at time 0. We have already set $r_n = rT/n$. Thus the price of the risk-free security at the last moment of time $t_n = T$ will be

$$\left(1 + \frac{rT}{n}\right)^n \approx e^{rT}$$

for large n.

How can we define the probability π and the model parameters u and d so as to obtain in the limit the given values for the drift μ and the volatility σ? Following

Cox, Ross and Rubinstein (the authors of the binomial model), this is done as follows. Define

$$\Delta = \frac{T}{n},$$

$$u = e^{\sigma\sqrt{\Delta}},\ d = e^{-\sigma\sqrt{\Delta}}, \tag{15.1}$$

$$\pi = \frac{1}{2}\left(1 + \frac{\mu}{\sigma}\sqrt{\Delta}\right). \tag{15.2}$$

Drift in the Binomial Model In the binomial model under consideration, the price $S(t_k) = S_k$ can be represented as $S_k = S_0 Z_1 \dots Z_k$, where Z_j are independent identically distributed random variables. The distribution of each Z_j coincides with the distribution of a random variable Z such that

$$Z = \begin{cases} u \text{ with probability } \pi, \\ d \text{ with probability } 1 - \pi. \end{cases}$$

Consequently,

$$\begin{aligned} E \ln \frac{S_{k+m}}{S_k} &= E \ln \frac{S_0 Z_1 \dots Z_{k+m}}{S_0 Z_1 \dots Z_k} \\ &= E \ln(Z_{k+1} \dots Z_{k+m}) \\ &= E(\ln Z_{k+1} + \dots + \ln Z_{k+m}) \\ &= m\, E \ln Z. \end{aligned}$$

Using (15.1) and (15.2), one finds that $\ln Z$ takes on the values $\ln u = \sigma\sqrt{\Delta}$ and $\ln d = -\sigma\sqrt{\Delta}$ with probabilities π and $(1 - \pi)$. This yields

$$\begin{aligned} E \ln Z &= \pi\, \sigma\sqrt{\Delta} - (1 - \pi)\sigma\sqrt{\Delta} \\ &= -\sigma\sqrt{\Delta} + 2\pi\sigma\sqrt{\Delta} \\ &= -\sigma\sqrt{\Delta} + (1 + \frac{\mu}{\sigma}\sqrt{\Delta})\sigma\sqrt{\Delta} \\ &= \mu\Delta. \end{aligned}$$

From this we conclude that

$$E \ln \frac{S(t_k + m\Delta)}{S(t_k)} = E \ln \frac{S_{k+m}}{S_k} = \mu\, m\Delta,$$

and so, *in the binomial model at hand, the price process $S(t_k)$ has the constant drift μ.*

Volatility in the Binomial Model We have

$$Var \ln \frac{S_{k+m}}{S_k} = Var[\ln Z_{k+1} + \ldots + \ln Z_{k+m}] = m\, Var \ln Z,$$

because Z_k are independent and have the same distribution as the random variable

$$Z = \begin{cases} u \text{ with probability } \pi, \\ d \text{ with probability } 1 - \pi. \end{cases}$$

Let us consider the case when $\Delta = T/n$ is small (or equivalently, n is large). Recalling the definition of π in (15.2), one finds that for small Δ

$$\pi \approx \frac{1}{2}. \tag{15.3}$$

We have [see (15.1)] $\ln u = \sigma\sqrt{\Delta}$, $\ln d = -\sigma\sqrt{\Delta}$, and so

$$E \ln Z \approx \frac{1}{2} \cdot \sigma\sqrt{\Delta} + \frac{1}{2} \cdot (-\sigma\sqrt{\Delta}) = 0$$

for small Δ (or equivalently, large n). Then

$$Var \ln Z \approx E(\ln Z)^2 \approx \frac{1}{2} \cdot [\sigma\sqrt{\Delta}]^2 + \frac{1}{2} \cdot [-\sigma\sqrt{\Delta}]^2 = \sigma^2 \Delta.$$

From this and from (15.3) we conclude that

$$Var \ln \frac{S(t_k + m\Delta)}{S(t_k)} = Var \ln \frac{S_{k+m}}{S_k} \approx \sigma^2 m\Delta,$$

where the approximation becomes accurate as $n \to \infty$ (or $\Delta = T/n \to 0$). Thus, *in the limit, the binomial price process has the constant volatility* σ.

15.4 Derivation of the Black–Scholes Formula

The Binomial Approximation of the Price of a Derivative Suppose we have to price a derivative security with a payoff function $F(S(T))$ (e.g., a European option), knowing that the nominal interest rate is r, and that the price process $S(t)$ for the underlying security has constant drift μ and volatility σ. We proceed as follows. We approximate the price process $S(t)$ by using the binomial model with the price process $S_k = S(t_k)$, $k = 0, \ldots, n$, described above and compute the price of the derivative within the binomial model. Then, by passing to the limit as the binomial approximation becomes more and more accurate, we obtain the sought-for price of the derivative.

The risk-neutral price of the derivative with the payoff $F(S_n)$ in the binomial model is

$$\Pi_n = (1 + r_n)^{-n} E^Q F(S_n), \tag{15.4}$$

where $Q(\omega)$ is the (unique) risk-neutral measure on the set Ω of market histories $\omega = (a_1, a_2, \dots, a_n)$. If ω is distributed according to Q, then the states of the world a_k are i.i.d., $P\{a_k = u\} = p$, and $P\{a_k = d\} = 1 - p$, where

$$p = \frac{1 + r_n - d}{u - d} \quad \left[r_n = \frac{rT}{n} = r\Delta \right].$$

Since $S_n = S_0 Z_1 Z_2 \dots Z_n$ and $Z_k(a) = a$, formula (15.4) can be written

$$\Pi_n = (1 + r_n)^{-n} E^Q F(S_0 Z_1 \dots Z_n),$$

where Z_k are i.i.d., $Z_k = u$ with probability p and $Z_k = d$ with probability $1 - p$.

Application of the Central Limit Theorem We have obtained the following formula for the price Π_n:

$$\Pi_n = (1 + r_n)^{-n} E^Q F(S_0 Z_1 \dots Z_n), \tag{15.5}$$

where Z_k are i.i.d., $Z_k = u$ with probability p and $Z_k = d$ with probability $1 - p$. It is convenient to replace Z_k by $V_k = \ln Z_k$. Then $Z_k = e^{V_k}$, and so

$$E^Q F(S_0 Z_1 \dots Z_n) = E^Q F(S_0 e^{V_1 + \dots + V_n}),$$

where each of V_k takes on the values $\ln u = \sigma\sqrt{\Delta}$ and $\ln d = -\sigma\sqrt{\Delta}$ with probabilities p and $1 - p$.

We will use the **Central Limit Theorem of Probability Theory** which, stated informally (see any text on Probability for a thorough treatment), says the following:

If a random variable W is the sum

$$W = V_1 + \dots + V_n$$

of a large number n of small i.i.d. random variables $V_1, \dots, V_n$, then the distribution of W can be approximated by the distribution of a normal (Gaussian) random variable with the expectation EW and the variance $Var\, W$.

One can formulate this result rigorously and prove that it holds in the current context. It implies that formula (15.5) leads in the limit to the following formula for the price of the derivative security with the payoff function $F(S(T))$:

$$\Pi = e^{-rT} EF(S_0 e^W),$$

where W is a Gaussian random variable, the expectation and the variance of which we will now compute. [Note: e^{-rT} is the limit of $(1 + r_n)^{-n}$ in (15.5).]

Analyzing the Sum $W = V_1 + \ldots + V_n$ Recall that $W = V_1 + \ldots + V_n$, where $V_1, \ldots, V_k$ are i.i.d. and each is distributed as the random variable

$$V = \begin{cases} \ln u = \sigma\sqrt{\Delta} & \text{with probability } p, \\ \ln d = -\sigma\sqrt{\Delta} & \text{with probability } 1-p, \end{cases}$$

$$p = \frac{1 + r\Delta - d}{u - d}, \tag{15.6}$$

and $\Delta = T/n$. The variance (resp. the expectation) of a sum of i.i.d. random variables is equal to the sum of their variances (resp. their expectations). Thus

$$EW = n\, EV \text{ and } Var\, W = n\, Var\, V.$$

By using the approximate formulas[2]

$$u \approx 1 + \sigma\sqrt{\Delta} + \frac{\sigma^2\Delta}{2}, \quad d \approx 1 - \sigma\sqrt{\Delta} + \frac{\sigma^2\Delta}{2}, \tag{15.7}$$

and substituting these approximate values of u and d into (15.6), we get

$$p \approx \frac{\sigma\sqrt{\Delta} - \frac{\sigma^2\Delta}{2} + r\Delta}{2\sigma\sqrt{\Delta}} = \frac{1}{2} - \frac{\sigma\sqrt{\Delta}}{4} + \frac{r\sqrt{\Delta}}{2\sigma}. \tag{15.8}$$

Computing EW The expectation of a random variable taking on two values a and b with probabilities p and $1-p$ is equal to $pa + (1-p)b = b + p(a-b)$. By applying this elementary formula to the random variable V which takes on the

[2]Formulas (15.7) for $u = e^{\sigma\sqrt{\Delta}}$ and $d = e^{-\sigma\sqrt{\Delta}}$ are obtained by using the approximate formula for the function e^x:

$$e^x \approx 1 + x + \frac{x^2}{2}.$$

The precise formula is $e^x = 1 + x + \frac{x^2}{2} + \ldots + \frac{x^m}{m!} + \ldots$.

values $a = \sigma\sqrt{\Delta}$ and $b = -\sigma\sqrt{\Delta}$ and using the approximate expression for p obtained above [see (15.8)], we find:

$$\begin{aligned} EV &= -\sigma\sqrt{\Delta} + \left(\frac{1}{2} - \frac{\sigma\sqrt{\Delta}}{4} + \frac{r\sqrt{\Delta}}{2\sigma}\right) 2\sigma\sqrt{\Delta} \\ &= -\sigma\sqrt{\Delta} + \sigma\sqrt{\Delta} - \frac{\sigma^2\Delta}{2} + r\Delta \\ &= -\frac{\sigma^2\Delta}{2} + r\Delta \\ &= \left(-\frac{\sigma^2}{2} + r\right)\frac{T}{n} \end{aligned}$$

(recall that $\Delta = T/n$), and so

$$EW = nEV = \left(r - \frac{\sigma^2}{2}\right)T.$$

Computing *Var W* The variance of a random variable taking on two values a and b with probabilities p and $1 - p$ is equal to $p(1-p)(a-b)^2$. By applying this formula to V and observing that $p \approx 1/2$ for small $\Delta = T/n$ [see (15.8)], we get

$$Var\, V \approx p \cdot (1-p) \cdot \left(\sigma\sqrt{\Delta} + \sigma\sqrt{\Delta}\right)^2 = \sigma^2\Delta = \frac{\sigma^2 T}{n},$$

which in the limit yields

$$Var\, W = n\, Var\, V = \sigma^2 T.$$

A Pricing Formula for an American Derivative with Payoff *F* We have obtained the following result.

Theorem 15.1 *If the price process $S(t)$ of the underlying security has a constant drift and a constant volatility σ, then the price of a derivative security with payoff function $F(S(T))$ is given by the formula*

$$\Pi = e^{-rT} EF\left(S(0)e^W\right), \tag{15.9}$$

where r is the nominal interest rate and W is a Gaussian random variable with mean

$$EW = \left(r - \frac{\sigma^2}{2}\right)T$$

and variance

$$Var\, W = \sigma^2 T.$$

Recall that a *Gaussian (normal)* random variable W is a random variable having the following density function:

$$\phi(w) = \frac{1}{\sqrt{2\pi s^2}} \exp\left[-\frac{(w-m)^2}{2s^2}\right],$$

where $m = EW$ and $s^2 = Var\, W$. By using this formula for the density $\phi(x)$, we can write (15.9) as

$$\Pi = \frac{e^{-rT}}{\sqrt{2\pi s^2}} \int_{-\infty}^{+\infty} F(S(0)e^w) \exp\left[-\frac{(w-m)^2}{2s^2}\right] dw,$$

where $s^2 = \sigma^2 T$ and $m = (r - \sigma^2/2)T$. This yields an explicit analytic expression for Π.

Remark It is important to note that formula (15.9) does not include μ—the drift of the process $S(t)$, and so the price Π depends only on σ, r, T and $S(0)$!

The Black–Scholes Formula We can use Theorem 15.1, for example, in the case where the derivative security is the European call option with strike price K, i.e., $F(S(T)) = \max\{S(T) - K, 0\}$. This allows us to obtain quite easily the Black–Scholes formula. As before, suppose that the price process $S(t)$ for the underlying security has constant drift and constant volatility σ. Let r be the nominal interest rate.

Theorem 15.2 *The price of the European call option with strike price K is given by the formula*

$$\Pi = e^{-rT} E \max\{S(0)e^W - K, 0\}, \tag{15.10}$$

where W is a Gaussian random variable with mean

$$EW = \left(r - \frac{\sigma^2}{2}\right) T \tag{15.11}$$

and variance

$$Var\, W = \sigma^2 T. \tag{15.12}$$

An explicit formula for Π can be written as

$$\Pi = S(0)\Phi(z) - Ke^{-rT}\Phi(z - \sigma\sqrt{T}) \tag{15.13}$$

where

$$z = \frac{rT + \sigma^2 T/2 - \ln(K/S(0))}{\sigma\sqrt{T}}.$$

Here $\Phi(z)$ is the distribution function of the standard Gaussian random variable with mean 0 and variance 1:

$$\Phi(z) = \frac{1}{\sqrt{2\pi}}\int_{-\infty}^{z} \exp\left(-\frac{w^2}{2}\right) dw.$$

Formula (15.13) is the **Black–Scholes formula**.

16 Problems and Exercises II

Question 16.1 (Put-Call Parity) By using the dynamic securities market model, prove that under hypothesis (**NA**), the no arbitrage price π^c of the European call option with strike price K and the no arbitrage price π^p of the European put option with strike price K are related to each other by the formula (*put-call parity*):

$$\pi^c - \pi^p + \frac{K}{(1+r)^T} = S_0, \tag{16.1}$$

where r is the risk-free interest rate and S_0 is the price of the underlying security at time 0. You can assume that the market is complete and does not allow for arbitrage opportunities.

Answer According to the Fundamental Theorem of Asset Pricing, the no-arbitrage hypothesis implies the existence of a risk-neutral probability measure Q. Since the market is complete, the contingent claim $F(S_T)$ is hedgeable. Then the no-arbitrage price of a derivative security with the payoff $F(S_T)$ is equal to

$$\frac{1}{(1+r)^T} E^Q F(S_T), \tag{16.2}$$

where E^Q is the expectation with respect to the risk-neutral measure Q. The risk-neutrality of Q implies

$$S_0 = \frac{1}{(1+r)^T} E^Q S_T. \tag{16.3}$$

I.V. Evstigneev et al., *Mathematical Financial Economics*, Springer Texts in Business and Economics, DOI 10.1007/978-3-319-16571-4_16

By virtue of (16.2), we have

$$\pi^c = \frac{1}{(1+r)^T} E^Q \max\{S_T - K, 0\}, \tag{16.4}$$

$$\pi^p = \frac{1}{(1+r)^T} E^Q \max\{K - S_T, 0\}. \tag{16.5}$$

Subtracting (16.5) from (16.4) and using the formula

$$\max\{S_T - K, 0\} - \max\{K - S_T, 0\} = S_T - K$$

(verify it!), we get

$$\pi^c - \pi^p = \frac{E^Q S_T - K}{(1+r)^T},$$

which, combined with (16.3), yields (16.1).

Question 16.2 (European Put/Call) In the binomial model with parameters

$$T = 3,\ r = 0.1,\ u = 1.4,\ d = 0.8,\ S_0 = 1,$$

compute the price of the following derivative securities:

(i) the European put option with strike price $K = 0.9$;
(ii) the European call option with strike price $K = 1.5$.

Answer We start with computing the probability p:

$$p = \frac{1+r-d}{u-d} = \frac{1.1-0.8}{1.4-0.8} = \frac{0.3}{0.6} = \frac{1}{2}.$$

There are $8 = 2^3$ possible histories $\omega = (a_1, a_2, a_3)$ in the 3-period model; they all have the same probabilities: $1/8$. The price $S_T(\omega)$ can take four different values—see the table below (note that $S_0 = 1$). We compute these values and the corresponding values of the payoff functions of the two options (Table 16.1).

Since $u = 1.4$ and $d = 0.8$, one has

Therefore

$$E^Q \max\{0.9 - S_T, 0\} = \frac{3}{8} \cdot 0.004 + \frac{1}{8} \cdot 0.388 = 0.0015 + 0.0485 = 0.05,$$

$$\text{price of put} = \frac{1}{(1+r)^3} \cdot 0.05 = \frac{0.05}{(1.1)^3} = \frac{0.05}{1.331} = 0.038 \approx 0.04;$$

Table 16.1 Question 16.2: market scenarios, their probabilities and the payoffs

ω	$S_T(\omega)$	Prob.	$\max\{0.9 - S_T, 0\}$	$\max\{S_T - 1.5, 0\}$
(u,u,u)	$u^3 = 2.744$	1/8	0	1.244
$(u,u,d),(u,d,u),(d,u,u)$	$u^2d = 1.568$	3/8	0	0.068
$(d,d,u),(d,u,d),(u,d,d)$	$ud^2 = 0.896$	3/8	0.004	0
(d,d,d)	$d^3 = 0.512$	1/8	0.388	0

$$E^Q \max\{S_T - 1.5, 0\} = \frac{1}{8} \cdot 1.244 + \frac{3}{8} \cdot 0.068 = 0.1555 + 0.0255 = 0.181,$$

$$\text{price of call} = \frac{1}{(1+r)^3} \cdot 0.181 = \frac{0.181}{1.331} \approx 0.14.$$

Question 16.3 (Asian Put/Call) In the binomial model with parameters

$$T = 2,\ r = 0.1,\ u = 2.1,\ d = 0.7,\ S_0 = 300,$$

calculate the price of the *Asian call option* giving the right to buy the underlying security at time T at the price which is equal to the average of the prices $S_0, \ldots, S_T$:

$$\bar{S} = \frac{1}{T+1} \sum_{t=0}^{T} S_t.$$

Compute the price of the *Asian put option* giving the right to sell the underlying security at time T at the price $\bar{S}$. The payoff functions of these derivative securities are

$$\max\{S_T - \bar{S}, 0\} \text{ and } \max\{\bar{S} - S_T, 0\},$$

respectively.

Answer We start with computing the probability p:

$$p = \frac{1+r-d}{u-d} = \frac{1.1 - 0.7}{2.1 - 0.7} = \frac{0.4}{1.4} = \frac{2}{7}.$$

The results of further computations are presented below (see Tables 16.2 and 16.3).

$$u = 2.1,\ d = 0.7,$$

$$E^Q \max\{S_2 - \bar{S}, 0\} = \frac{4}{49} \cdot 572 + \frac{10}{49} \cdot 124 =$$

$$\frac{4 \cdot 572 + 1240}{49} = 72.$$

Table 16.2 Question 16.3: market scenarios and the corresponding prices

ω	S_0	S_1	S_2	$Q(\omega)$
(u, u)	300	630	1,323	4/49
(u, d)	300	630	441	10/49
(d, u)	300	210	441	10/49
(d, d)	300	210	147	25/49

Table 16.3 Question 16.3: derivative payoffs for all market scenarios

ω	$\bar{S}$	$\max\{S_2 - \bar{S}, 0\}$	$\max\{\bar{S} - S_2, 0\}$
(u, u)	751	572	0
(u, d)	457	0	16
(d, u)	317	124	0
(d, d)	219	0	72

$$E^Q \max\{\bar{S} - S_2, 0\} = \frac{10}{49} \cdot 16 + \frac{25}{49} \cdot 72 =$$

$$\frac{160 + 25 \cdot 72}{49} = 40.$$

$$\text{Price of Asian call} = \frac{72}{(1+r)^2} = \frac{72}{1.21} = 59.50.$$

$$\text{Price of Asian put} = \frac{40}{1.21} = 33.06.$$

Question 16.4 (Lookback Put/Call) The *lookback call option* allows an investor to buy the underlying asset at time T at the smallest of the prices $S_0, \ldots, S_T$. The *lookback put option* gives the right to sell the underlying asset at the greatest of the prices $S_0, \ldots, S_T$. The payoff functions of the derivative securities are

$$S_T - \min_{0 \le t \le T} S_t \text{ and } \max_{0 \le t \le T} S_t - S_T,$$

respectively. Compute the prices of these derivative securities in the binomial model with the parameters specified in the previous question.

Answer We use the same table as in the previous question:

Then we compute for each ω the values of the payoff functions for the lookback call and put options (see Tables 16.4 and 16.5).

$$E^Q(S_2 - \min S_t) = \frac{4}{49} \cdot 1023 + \frac{10}{49} \cdot 141 + \frac{10}{49} \cdot 231 =$$

$$\frac{4 \cdot 1023 + 1410 + 2310}{49} = 159.4286.$$

Table 16.4 Question 16.4: market histories, prices and probabilities

ω	S_0	S_1	S_2	$Q(\omega)$
(u, u)	300	630	1,323	$4/49$
(u, d)	300	630	441	$10/49$
(d, u)	300	210	441	$10/49$
(d, d)	300	210	147	$25/49$

Table 16.5 Question 16.4: derivative payoffs

ω	$\min S_t$	$\max S_t$	$S_2 - \min S_t$	$(\max S_t) - S_2$
(u, u)	300	1,323	1,023	0
(u, d)	300	630	141	189
(d, u)	210	441	231	0
(d, d)	147	300	0	153

$$E^Q[(\max S_t) - S_2] = \frac{10}{49} \cdot 189 + \frac{25}{49} \cdot 153 =$$

$$\frac{1890 + 25 \cdot 153}{49} = 116.6327.$$

$$\text{Price of lookback call} = \frac{159.4286}{(1+r)^2} = \frac{159.4286}{1.21} = 131.76.$$

$$\text{Price of lookback put} = \frac{116.6327}{1.21} = 96.39.$$

Question 16.5 (American Put) In the binomial model with parameters

$$T = 3,\ r = 0.1,\ u = 1.4,\ d = 0.8,\ S_0 = 1,$$

compute the upper price of the American put option with payoff function

$$F(S) = \max\{K - S, 0\},$$

where $K = 0.9$.

Answer We begin with computing

$$S_{t,j} = S_0 u^j d^{t-j},\ j = 0, 1, \ldots, t,$$

first for $t = 0$, then for $t = 1$, and so on (see Table 16.6).

Then we find $(1 + r)^t, t = 1, 2, 3$:

$$1 + r = 1.1,\ (1 + r)^2 = 1.21,\ (1 + r)^3 = 1.331.$$

Table 16.6 Question 16.5: prices in the binomial model

$S_{t,j}$	$t=0$	$t=1$	$t=2$	$t=3$
$j=0$	1	0.8	0.64	0.512
$j=1$		1.4	1.12	0.896
$j=2$			1.96	1.568
$j=3$				2.744

Table 16.7 Question 16.5: discounted payoffs

$F_t(S_{t,j})$	$t=0$	$t=1$	$t=2$	$t=3$
$j=0$	0	0.09	0.215	0.292
$j=1$		0	0	0.003
$j=2$			0	0
$j=3$				0

Table 16.8 Question 16.5: conditional values

$V_t(j)$	$t=0$	$t=1$	$t=2$	$t=3$
$j=0$	0.054	0.108	0.215	0.292
$j=1$		0.00075	0.0015	0.003
$j=2$			0	0
$j=3$				0

After that we compute

$$F_t(S_{t,j}) = \frac{(K - S_{t,j})^+}{(1+r)^t} = \frac{(0.9 - S_{t,j})^+}{(1.1)^t},$$

where $K = 0.9$, see Table 16.7. Then we calculate

$$p = \frac{1+r-d}{u-d} = \frac{1.1-0.8}{1.4-0.8} = \frac{0.3}{0.6} = 0.5,$$

and then compute $V_t(j)$ for $t = 3, 2, 1, 0$ by using the formulas

$$V_T(j) = F_T(S_{T,j}),\ j = 0, 1, \ldots, T;$$

$$V_t(j) = \max\left\{ F_t(S_{t,j}),\ \frac{V_{t+1}(j+1) + V_{t+1}(j)}{2} \right\}.$$

The results of the computations are presented in the table (Table 16.8):

Thus the upper price of the American put option is $V_0(0) = 0.054$.

Question 16.6 (American Barrier Put) This question deals with *barrier options*, that are exercised only if the price of the underlying security lies within certain limits—barriers (this restricts the liability of the seller). Consider the *American* *barrier put option* with strike price K that is exercised only if the current price of the underlying security is not less than some given level $B < K$. The payoff function of this option is as follows:

$$F(S) = \begin{cases} K - S & \text{if } B \leq S \leq K, \\ 0 & \text{otherwise.} \end{cases} \tag{16.6}$$

Compute the upper price of the American barrier put option [whose payoff function is given by formula (16.6)] in the binomial model with parameters

$$T = 3,\ r = 0,\ u = 1.6,\ d = 0.6,\ S_0 = 1000,$$

assuming that $K = 970$ and $B = 370$.

Answer We use the algorithm for computing the upper price of any American derivative security with a payoff function $F(S)$. For each $t = 0, 1, 2, \ldots, T$, we calculate the $t + 1$ possible values

$$S_{t,j} = S_0 u^j d^{t-j},\ j = 0, 1, \ldots, t\,,$$

of the price S_t. We define

$$F_t(S) = \frac{F(S)}{(1+r)^t}.$$

We compute by backward induction (from $t + 1$ to t) the numbers $V_t(j)$ ($j = 0, 1, 2, \ldots, t$) defined by the formulas

$$V_T(j) = F_T(S_{T,j}),\ j = 0, 1, \ldots, T,$$
$$V_t(j) = \max\{F_t(S_{t,j}),\ pV_{t+1}(j+1) + (1-p)V_{t+1}(j)\},$$

where

$$p = \frac{1 + r - d}{u - d}.$$

The number $V_0(0)$ is equal to the sought-for price.

We begin with computing

$$S_{t,j} = S_0 u^j d^{t-j},\ j = 0, 1, \ldots, t.$$

Table 16.9 Question 16.6: prices

$S_{t,j}$	$t=0$	$t=1$	$t=2$	$t=3$
$j=0$	1,000	600	360	216
$j=1$		1,600	960	576
$j=2$			2,560	1,536
$j=3$				4,096

Table 16.10 Question 16.6: zero discounted payoffs

$F_t(S_{t,j})$	$t=0$	$t=1$	$t=2$	$t=3$
$j=0$	0		0	0
$j=1$		0		
$j=2$			0	0
$j=3$				0

Table 16.11 Question 16.6: all discounted payoffs

$F_t(S_{t,j})$	$t=0$	$t=1$	$t=2$	$t=3$
$j=0$	0	370	0	0
$j=1$		0	10	394
$j=2$			0	0
$j=3$				0

The results are recorded in Table 16.9.

In the example at hand, we have $r = 0$, so that

$$F_t(S_{t,j}) = \frac{F(S_{t,j})}{(1+r)^t} = F(S_{t,j}).$$

According to the definition of the payoff function $F(S_{t,j})$, it is not equal to zero only if $370 \leq S_{t,j} \leq 970$. If these conditions are not satisfied, then $F(S_{t,j}) = 0$, which makes it possible to immediately fill most of the cells in the table of values of $F(S_{t,j})$, see Table 16.10. We fill the remaining cells by using the formula $F(S_{t,j}) = 970 - S_{t,j}$ (when $370 \leq S_{t,j} \leq 970$), see Table 16.11. Then we find

$$p = \frac{1+0-d}{u-d} = \frac{1-0.6}{1.6-0.6} = \frac{0.4}{1.0} = 0.4,$$

and then compute $V_t(j)$ for $t = 3, 2, 1, 0$ by using the formulas

$$V_T(j) = F(S_{T,j}),\ j = 0, 1, \ldots, T;$$
$$V_t(j) = \max\{F(S_{t,j}),\ 0.4 \cdot V_{t+1}(j+1) + 0.6 \cdot V_{t+1}(j)\}.$$

Table 16.12 Question 16.6: conditional values

$V_t(j)$	$t = 0$	$t = 1$	$t = 2$	$t = 3$
$j = 0$	278.74	370	157.6	0
$j = 1$		141.84	236.4	394
$j = 2$			0	0
$j = 3$				0

The results of the computations are presented in Table 16.12.
Thus the upper price of the American barrier put option is

$$V_0(0) = 278.74.$$

Part III

Growth and Equilibrium

17 Capital Growth Theory

17.1 Growth-Optimal Investments

Model Description We examine the following multiperiod investment problem: starting from some initial wealth available at time 0, find a self-financing trading strategy that maximizes the long-run growth rate of the investor's wealth. To examine this problem, we consider a version of the dynamic securities market model considered in the previous chapters.

- Trading on the market is possible at any of the dates: $t = 0, 1, 2, 3, \ldots$ (so that we deal here with an *infinite* time horizon).
- A finite set

$$A = \{a^1, \ldots, a^L\}$$

is given, elements of which are interpreted as possible *states of the world*. The state of the world which is realized at time $t = 1, 2, \ldots$ is denoted by a_t.
- The states of the world $a_1, a_2, \ldots$ are independent and identically distributed (i.i.d.). The distribution of each a_t is specified by L numbers (as many as there are states of the world)

$$p(a^1) > 0, p(a^2) > 0, \ldots, p(a^L) > 0$$

such that the probability that $a_t = a^l$ is equal to $p(a^l)$; in symbols,

$$P\{a_t = a^l\} = p(a^l).$$

The sum $\sum_{l=1}^{L} p(a^l)$ is equal to 1.

I.V. Evstigneev et al., *Mathematical Financial Economics*, Springer Texts in Business and Economics, DOI 10.1007/978-3-319-16571-4_17

- There are N *securities (assets)* $i = 1, \dots, N$. Each asset is characterized by its *return* $R^i(a)$, depending on the state of the world a. If at date t the state of the world is a_t, then the return on asset i is

$$R_t^i = R^i(a_t).$$

Since the states of the world $a_1, a_2, \dots$ are i.i.d., the vectors of asset returns

$$R_t = (R_t^1, \dots, R_t^N),\ t = 1, 2, \dots,$$

are i.i.d. as well.

Asset Returns and Prices The return R_t^i on asset i can be expressed through its prices S_{t-1}^i and S_t^i as

$$R_t^i = \frac{S_t^i - S_{t-1}^i}{S_{t-1}^i}.$$

This yields: $S_{t-1}^i R_t^i = S_t^i - S_{t-1}^i$ and

$$S_t^i = (1 + R_t^i) S_{t-1}^i.$$

Thus we obtain

$$S_t^i = (1 + R_t^i) \dots (1 + R_1^i) S_0^i. \tag{17.1}$$

Since the random variables $1 + R_t^i = 1 + R^i(a_t)$ $(t = 1, 2, \dots)$ are independent (and S_0^i is a constant), the price process S_t^i of each asset i is a *geometric random walk*. This is an assumption commonly accepted in finance. The vector of asset *prices* at time $t = 0, 1, \dots$ is denoted by $S_t = (S_t^1, \dots, S_t^N)$. The vector S_t $(t \geq 1)$ is a function of the *partial history* $\omega^t = (a_1, \dots, a_t)$ in view of formula (17.1) where $R_t^i = R^i(a_t)$. We assume that $S_0^i > 0$ and $R_t^i > -1$, which implies $S_t^i > 0$.

Trading Strategies A *trading strategy* (*investment strategy*) H is a sequence

$$H = (h_0, h_1, \dots)$$

of investor's portfolios specified for each moment of time $t = 0, 1, \dots$. The portfolio h_0 is fixed (non-random). For each $t \geq 1$, the portfolio h_t depends on $\omega^t = (a_1, \dots, a_t)$:

$$h_t = h_t(\omega^t),\ t = 1, 2, \dots$$

The coordinate h_t^i of the vector $h_t = (h_t^1, \ldots, h_t^N)$ indicates the number of units of asset i in the portfolio h_t. We will focus on the analysis *self-financing* strategies, i.e. those which satisfy

$$\langle S_t, h_{t-1} \rangle = \langle S_t, h_t \rangle, \ t = 1, 2, \ldots$$

When analyzing the problem of growth optimal investments, we will assume that all portfolio positions are *non-negative*. This requirement will be included into the definition of a trading strategy. Thus, for each trading strategy $H = (h_0, h_1, \ldots)$, we have $h_0 \geq 0$ and $h_t(\omega^t) \geq 0$ for all $t \geq 1$ and all ω^t.

For a trading strategy $H = (h_0, h_1, \ldots)$, we will denote by V_t^H the value of the portfolio h_t expressed in terms of the price vector S_t:

$$V_t^H = \langle S_t, h_t \rangle = S_t^1 h_t^1 + S_t^2 h_t^2 + \ldots + S_t^N h_t^N.$$

17.2 Strategies in Terms of Investment Proportions

Investment Proportions To study the problem of growth optimal investments, it will be convenient to specify self-financing trading strategies in terms of *investment proportions*. Consider the set of N-dimensional vectors $x = (x^1, \ldots, x^N)$ such that

$$x^i \geq 0 \text{ and } \sum x^i = 1.$$

This set is the N-dimensional unit *simplex*. Its elements are interpreted as vectors of *proportions*.

Suppose we are given a sequence of vectors

$$x_0, x_1(\omega^1), x_2(\omega^2), x_3(\omega^3), \ldots$$

with values in the unit simplex, where the vector

$$x_t(\omega^t) = (x_t^1(\omega^t), \ldots, x_t^N(\omega^t))$$

depends on the market history $\omega^t = (a_1, \ldots, a_t)$. Given a number $w > 0$ (initial wealth) and the above sequence, we can construct recursively a trading strategy $H = (h_0, h_1, \ldots)$ as follows. Define the initial portfolio $h_0 = (h_0^1, \ldots, h_0^N)$ by

$$h_0^i = \frac{x_0^i w}{S_0^i}, \ i = 1, 2, \ldots, N. \tag{17.2}$$

If the portfolios $h_0, \ldots, h_{t-1}$ are already defined, we define the portfolio $h_t = (h_t^1, \ldots, h_t^N)$ by

$$h_t^i = \frac{x_t^i \langle S_t, h_{t-1} \rangle}{S_t^i}, \; i = 1, 2, \ldots, N. \tag{17.3}$$

For $t \geq 1$, the portfolio h_t is, generally, a function of ω^t because x_t^i and S_t are functions of ω^t, so that we deal here with a sequence of contingent portfolios.

Formulas (17.2) and (17.3) by which we defined the trading strategy H can be written as follows

$$S_0^i h_0^i = x_0^i w, \; S_t^i h_t^i = x_t^i \langle S_t, h_{t-1} \rangle, \; i = 1, 2, \ldots, N. \tag{17.4}$$

By summing up over $i = 1, 2, \ldots, N$, we get

$$\langle S_0, h_0 \rangle = \sum S_0^i h_0^i = w, \; \langle S_t, h_t \rangle = \sum S_t^i h_t^i = \langle S_t, h_{t-1} \rangle \tag{17.5}$$

because $\sum_i x_t^i = 1$ for all $t \geq 0$. This shows that the strategy H we have constructed is self-financing, and its initial wealth V_0^H is equal to w.

Furthermore, we find

$$S_0^i h_0^i = x_0^i \langle S_0, h_0 \rangle, \; S_t^i h_t^i = x_t^i \langle S_t, h_t \rangle.$$

Thus, the *fraction of investor's total wealth* $V_t^H = \langle S_t, h_t \rangle$ *at time* $t \geq 0$ *invested in asset* i *is equal to* x_t^i.

Remark 17.1 The above considerations show that self-financing trading strategies can be specified in terms of sequences $X = (x_0, x_1, \ldots)$ of vectors $x_t = x_t(\omega^t)$ of investments proportions. This is often more convenient than specifying such strategies in terms of portfolios h_t. The self-financing condition $\langle S_t, h_t \rangle = \langle S_t, h_{t-1} \rangle$ imposes a joint constraint on the pair of portfolios h_{t-1} and h_t, whereas the vectors x_{t-1} and x_t can be chosen independently of each other.

Expressing Wealth via Investment Proportions and Asset Returns Define

$$Z^i(a) = 1 + R^i(a)$$

and consider the vector $Z_t = (Z_t^1, \ldots, Z_t^N)$ with coordinates

$$Z_t^i = Z^i(a_t).$$

Observe that

$$Z_t^i = 1 + R_t^i = 1 + \frac{S_t^i - S_{t-1}^i}{S_{t-1}^i} = \frac{S_t^i}{S_{t-1}^i}.$$

Proposition 17.1 *If an investment strategy $H = (h_0, h_1, \ldots)$ with initial wealth w is defined by a sequence $x_0, x_1, \ldots$ of investment proportions, then the portfolio value V_t^H can be expressed by the formula*

$$V_T^H = w\langle x_0, Z_1\rangle\langle x_1, Z_2\rangle \ldots \langle x_{T-1}, Z_T\rangle.$$

Proof We have

$$V_t^H = \langle S_t, h_t\rangle = \langle S_t, h_{t-1}\rangle = \sum_{i=1}^{N} S_t^i h_{t-1}^i = \sum_{i=1}^{N} \frac{S_t^i}{S_{t-1}^i} S_{t-1}^i h_{t-1}^i$$

$$= \sum_{i=1}^{N} Z_t^i S_{t-1}^i h_{t-1}^i = \sum_{i=1}^{N} Z_t^i x_{t-1}^i V_{t-1}^H = V_{t-1}^H \sum_{i=1}^{N} Z_t^i x_{t-1}^i = V_{t-1}^H \langle x_{t-1}, Z_t\rangle,$$

which yields the result. □

17.3 Results for Simple Strategies

Growth Optimal Investments: Simple Strategies Let us consider those strategies for which the investment proportions are constant (independent of t and ω^t). Such strategies are called *simple*. Each of them is determined by a vector of proportions x in the unit simplex. We will examine the following problem: *find that vector of investment proportions for which the corresponding simple strategy exhibits growth of wealth faster than any other simple strategy.*

To formulate a solution to the problem consider that vector x^* in the unit simplex which maximizes the function

$$U(x) = E \ln\langle x, Z(a_t)\rangle$$

(since a_t are i.i.d., $U(x)$ does not depend on t). The vector x^* defines the *log-optimal proportions*. Assume x^* is *unique*. Consider the simple trading strategy H^* with the vector of proportions x^* and some initial wealth $w^* > 0$. A central result of this chapter is as follows:

Theorem 17.1 *Let H be a simple trading strategy with a vector of investment proportions $x \neq x^*$ and initial wealth $w > 0$. Then we have*

$$\frac{V_T^{H^*}}{V_T^H} \to \infty \text{ with probability one.}$$

Law of Large Numbers The proof of Theorem 17.1 is based on a fundamental law of Probability Theory—The Law of Large Numbers. It can be formulated as follows:

Let $\xi_1, \xi_2, \xi_3, \dots$ be independent identically distributed random variables having finite expectation $\mu = E\xi_t$. Then

$$\frac{\xi_1 + \xi_2 + \ldots + \xi_T}{T} \to \mu$$

with probability one.

Proof of Theorem 17.1 For $x \neq x^*$, we have $U(x) < U(x^*)$ because x^* is the unique vector maximizing $U(x) = E \ln\langle x, Z(a_t)\rangle$. Define

$$\xi_t = \ln \frac{\langle x^*, Z(a_t)\rangle}{\langle x, Z(a_t)\rangle} = \ln\langle x^*, Z(a_t)\rangle - \ln\langle x, Z(a_t)\rangle$$

and put $\mu = E\xi_t = U(x^*) - U(x)$. We know that

$$V_T^{H^*} = w^*\langle x^*, Z_1\rangle\langle x^*, Z_2\rangle \ldots \langle x^*, Z_T\rangle,$$
$$V_T^{H} = w\langle x, Z_1\rangle\langle x, Z_2\rangle \ldots \langle x, Z_T\rangle.$$

Consequently,

$$\frac{1}{T}\ln\frac{V_T^{H^*}}{V_T^H} = \frac{1}{T}\ln\frac{w^*}{w} + \frac{1}{T}\left(\ln\frac{\langle x^*, Z_1\rangle}{\langle x, Z_1\rangle} + \ldots + \ln\frac{\langle x^*, Z_T\rangle}{\langle x, Z_T\rangle}\right)$$
$$= \frac{1}{T}\ln\frac{w^*}{w} + \frac{1}{T}(\xi_1 + \ldots + \xi_T) \to \mu > 0$$

with probability one.

The fact that the sequence $\frac{1}{T}\ln[V_T^{H^*}/V_T^H]$ converges to $\mu > 0$ implies that

$$\frac{1}{T}\ln(V_T^{H^*}/V_T^H) > \mu/2 > 0 \tag{17.6}$$

for all T large enough, or equivalently,

$$V_T^{H^*}/V_T^H \geq e^{T(\mu/2)} \tag{17.7}$$

for all T large enough. It follows from (17.7) that

$$V_T^{H^*}/V_T^H \to \infty. \tag{17.8}$$

Since (17.6) and (17.7) hold with probability one (for all T large enough), (17.8) holds with probability one, which completes the proof. □

Horse Race Model Consider an illustrative example of application of the above theory. There is a series of repeated races with N horses. Only one horse wins in each race. The probability that horse i wins is $p(i) > 0$. If a gambler bets y Euros on a horse that wins, the amount gained is Wy ($W > 0$ is a fixed number). If the horse does not win, the amount gained is 0. Consider the following betting strategy. Fix some proportions $x^1, \ldots, x^N$ and distribute all wealth available after each race across horses $i = 1, 2, \ldots, N$ in the proportions $x^1, \ldots, x^N$.

Question: What proportions $x^1, \ldots, x^N$ will guarantee the highest growth rate of wealth in the long run?

Answer: The growth optimal proportions x^i are equal to the probabilities $p(i)$ (the *Kelly rule —"betting your beliefs"*).

To justify the answer we use the above results. Consider the state space $A = \{1, \ldots, N\}$. Let $a_t = i$ if horse i wins. The states of the world $a_1, a_2, \ldots$ are i.i.d. with probabilities $P\{a_t = i\} = p(i)$. Define

$$Z^i(a) = \begin{cases} W, & \text{if } a = i, \\ 0, & \text{otherwise.} \end{cases}$$

Suppose a gambler with initial wealth w uses the betting strategy defined by the vector of proportions $x = (x^1, \ldots, x^N)$. Then the gambler's wealth after T races is

$$w\langle x, Z(a_1)\rangle \ldots \langle x, Z(a_T)\rangle, \tag{17.9}$$

where the scalar product $\langle x, Z(a_t)\rangle = \sum_i x^i Z^i(a_t)$ is equal to $x^i W$ when $a_t = i$. We know that the maximum growth rate of (17.9) is attained for that x^* which maximizes

$$E \ln\langle x, Z(a_t)\rangle = \sum_i p(i) \ln\langle x, Z(i)\rangle = \sum_i p(i) \ln(x^i W)$$
$$= \ln W + \sum_i p(i) \ln x^i.$$

The fact that the maximum is attained at $x^* = (p(1), \ldots, p(N))$ follows from the elementary inequality

$$\sum_i p(i) \ln p(i) > \sum_i p(i) \ln x^i,$$

holding for all $x = (x^1, \ldots, x^N) \neq x^*$ in the unit simplex.

The above gambling scheme may be interpreted as a highly idealized model of a financial market. There are N assets and N states of the world. Asset

$i = 1, 2, \ldots, N$ pays W if the state i is realized and 0 otherwise (such assets are called *Arrow securities*). How to invest in order to achieve the fastest growth rate of wealth almost surely? As the above result shows, one has to allocate wealth in the same proportions as the probabilities of the states $i = 1, 2, \ldots, N$.

Capital Growth Theory: Continued 18

18.1 Log-Optimal Strategies

Log-Optimal Proportions We have seen that the vector of log-optimal proportions x^* plays a crucial role in growth optimal investments. Let us investigate its properties in more detail. Recall that x^* maximizes the function

$$U(x) = E \ln\langle x, Z(a_t)\rangle$$

on the unit simplex. Suppose for simplicity that $x^* > 0$. Then for any vector w with $\sum w^i = 0$ and any number γ with a sufficiently small absolute value $|\gamma|$ the point

$$x^* + \gamma w$$

belongs to the simplex, and the function

$$f(\gamma) = E \ln\langle x^* + \gamma w, Z(a_t)\rangle$$

attains its maximum at $\gamma = 0$. By differentiating

$$f(\gamma) = \sum_a p(a) \ln\langle x^* + \gamma w, Z(a)\rangle$$

with respect to γ, we get

$$f'(\gamma) = \sum_a p(a) \frac{\langle w, Z(a)\rangle}{\langle x^* + \gamma w, Z(a)\rangle} = E \frac{\langle w, Z(a_t)\rangle}{\langle x^* + \gamma w, Z(a_t)\rangle}.$$

I.V. Evstigneev et al., *Mathematical Financial Economics*, Springer Texts in Business and Economics, DOI 10.1007/978-3-319-16571-4_18

Since $f(\gamma)$ is concave, it attains its maximum at 0 if and only if

$$f'(0) = E\frac{\langle w, Z(a_t)\rangle}{\langle x^*, Z(a_t)\rangle} = 0.$$

Necessary and Sufficient Conditions for Log-Optimality Assuming $x^* > 0$, we obtained that x^* is a vector of log-optimal proportions if and only if

$$E\frac{\langle w, Z(a_t)\rangle}{\langle x^*, Z(a_t)\rangle} = 0$$

for any vector w with $\sum w^i = 0$. Denote by e_i the vector whose coordinates are 0 except the ith coordinate which is 1. Then for any i and j the sum of the coordinates of $w = e_i - e_j$ is equal to 0, and the above formula can be applied to this vector. Since $\langle e_i, Z(a)\rangle = Z^i(a)$, we find

$$E\frac{Z^i(a_t) - Z^j(a_t)}{\langle x^*, Z(a_t)\rangle} = 0 \text{ for all } i \text{ and } j,$$

and so

$$E\frac{Z^1(a_t)}{\langle x^*, Z(a_t)\rangle} = E\frac{Z^2(a_t)}{\langle x^*, Z(a_t)\rangle} = \ldots = E\frac{Z^N(a_t)}{\langle x^*, Z(a_t)\rangle}.$$

Denote each of these equal numbers by λ. Thus

$$E\frac{x^{*i} Z^i(a_t)}{\langle x^*, Z(a_t)\rangle} = \lambda x^{*i}.$$

By summing up over i, we find

$$\lambda = \sum \lambda x^{*i} = \sum E\frac{x^{*i} Z^i(a_t)}{\langle x^*, Z(a_t)\rangle} = E\frac{\langle x^*, Z(a_t)\rangle}{\langle x^*, Z(a_t)\rangle} = 1,$$

which yields

$$E\frac{Z^i(a_t)}{\langle x^*, Z(a_t)\rangle} = 1, \; i = 1, 2, \ldots, N.$$

These equalities are *necessary and sufficient conditions for the log-optimality of* $x^* > 0$.

Martingales Consider the process of random states of the world $a_1, a_2, \ldots$. (In our context these states are i.i.d., but more general processes can be considered.) A sequence of random variables (scalar- or vector-valued)

$$\xi_0, \xi_1(\omega^1), \xi_2(\omega^2), \ldots$$

having finite expectations $E|\xi_t|$ is called a *martingale* if

$$E(\xi_{t+1}|\omega^t) = \xi_t. \tag{18.1}$$

This equality means that the best prediction of the state ξ_{t+1} of the process $\xi_0, \xi_1, \xi_2, \ldots$ at time $t + 1$ ("tomorrow") based on the observation of the history $\omega^t = (a_1, \ldots, a_t)$ is ξ_t—the state of the process "today." The notion appeared in the theory of gambling. Martingales describe wealth dynamics in fair games.

If $\xi_0, \xi_1, \xi_2, \ldots$ are vector-valued random variables, $\xi_t = (\xi_t^1, \ldots, \xi_t^N)$, equality (18.1) is understood coordinatewise:

$$E(\xi_{t+1}^i|\omega^t) = \xi_t^i,\ i = 1, 2, \ldots, N.$$

18.2 Growth-Optimal and Numeraire Strategies

Numeraire Portfolio The term *numeraire portfolio* (Long 1990)[1] actually refers to some special trading strategy $H = (h_0, h_1, \ldots)$ defined by a sequence of contingent portfolios $h_0, h_1(\omega^1), h_2(\omega^2), \ldots$. A trading strategy $H = (h_0, h_1, \ldots)$ is called a *numeraire portfolio* (or a *numeraire strategy*) if the sequence

$$\xi_t = \frac{S_t}{V_t^H}$$

is a martingale. According to this definition, the asset price process denominated in terms of the numeraire portfolio value V_t^H forms a martingale. ("Numeraire" means a unit of measurement.)

Theorem 18.1 *Let x^* be the vector of log-optimal proportions and H^* the strategy generated by it (with some initial wealth $w^* > 0$). Then H^* is a numeraire strategy.*

Proof We know that

$$V_t^{H^*} = w^* \langle x^*, Z(a_1)\rangle \ldots \langle x^*, Z(a_t)\rangle.$$

[1] Long Jr., J. B., The numeraire portfolio, Journal of Financial Economics 26, 29–69, 1990.

Thus we need to prove the equality

$$E\left[\frac{S_{t+1}}{w^*\langle x^*, Z(a_1)\rangle \dots \langle x^*, Z(a_{t+1})\rangle} \mid \omega^t\right] = \frac{S_t}{w^*\langle x^*, Z(a_1)\rangle \dots \langle x^*, Z(a_t)\rangle}.$$

When computing the above conditional expectation, we fix $\omega^t = (a_1, \dots, a_t)$ and so the values of the variables $\langle x^*, Z(a_1)\rangle$, …, $\langle x^*, Z(a_t)\rangle$ become fixed ("non-random"). Therefore they can be cancelled out in the above equality, which reduces to the following one:

$$E\left[\frac{S^i_{t+1}}{\langle x^*, Z(a_{t+1})\rangle} \mid \omega^t\right] = S^i_t,\ i = 1, 2, \dots, N.$$

Since S^i_t is a function of ω^t, and ω^t is fixed, we can write this equality as

$$E\left[\frac{S^i_{t+1}/S^i_t}{\langle x^*, Z(a_{t+1})\rangle} \mid \omega^t\right] = 1. \tag{18.2}$$

Recall that $S^i_{t+1}/S^i_t = Z^i(a_{t+1})$, and so (18.2) can be written as

$$E\left[\frac{Z^i(a_{t+1})}{\langle x^*, Z(a_{t+1})\rangle} \mid \omega^t\right] = 1.$$

Here, a_{t+1} is independent of $\omega^t = (a_1, \dots, a_t)$, and therefore the conditional expectation reduces to the unconditional one, and the last equality becomes

$$E\frac{Z^i(a_{t+1})}{\langle x^*, Z(a_{t+1})\rangle} = 1.$$

But this is exactly the necessary and sufficient condition for the log-optimality of x^* we obtained above!

The proof is complete. □

18.3 Growth-Optimality for General Strategies

Comparing Simple and General Strategies Martingale properties associated with log-optimal investment proportions provide powerful tools for the analysis of growth optimal investments. We will demonstrate their use by comparing the performance of the growth optimal simple strategy H^* with the fixed vector of proportions x^* and any other strategy H generated by *any* sequence $x_0, x_1(\omega^1), x_2(\omega^2), \dots$ of vectors of proportions. Let $w^* > 0$ be the initial wealth of H^* and let $w > 0$ be the initial wealth of H.

Theorem 18.2 *Let V_t^H (resp. $V_t^{H^*}$) be the portfolio value at time t for the strategy H (resp. H^*). The sequence*

$$\eta_t = \frac{V_t^H}{V_t^{H*}}$$

has a finite limit with probability one.

This result means that the wealth of an investor using the strategy H cannot grow infinitely faster than the wealth of an investor using H^*. In this sense, no strategy H can outperform H^*.

Martingale Convergence Theorem The proof of Theorem 18.2 is based on the *martingale convergence theorem*, a fundamental principle of Probability Theory, which asserts: *Any non-negative martingale has a finite limit with probability one.*

Proof of Theorem 18.2 It is sufficient to show that the sequence

$$\eta_t = \frac{V_t^H}{V_t^{H*}}$$

is a martingale, i.e.,

$$E\left[\frac{w\langle x_0, Z(a_1)\rangle \dots \langle x_t, Z(a_{t+1})\rangle}{w^*\langle x^*, Z(a_1)\rangle \dots \langle x^*, Z(a_{t+1})\rangle} \mid \omega^t\right] = \frac{w\langle x_0, Z(a_1)\rangle \dots \langle x_{t-1}, Z(a_t)\rangle}{w^*\langle x^*, Z(a_1)\rangle \dots \langle x^*, Z(a_t)\rangle}. \tag{18.3}$$

By cancelling out those terms in this equality which depend only on $\omega^t = (a_1 \dots, a_t)$ and not on a_{t+1} (we fix them when computing the conditional expectation), we obtain that (18.3) is equivalent to

$$E\left[\frac{\langle x_t, Z(a_{t+1})\rangle}{\langle x^*, Z(a_{t+1})\rangle} \mid \omega^t\right] = 1.$$

Here x_t depends on ω^t (hence fixed), and so

$$E\left[\frac{\langle x_t, Z(a_{t+1})\rangle}{\langle x^*, Z(a_{t+1})\rangle} \mid \omega^t\right] = \sum_i x_t^i E\left[\frac{Z^i(a_{t+1})}{\langle x^*, Z(a_{t+1})\rangle} \mid \omega^t\right] = 1$$

because

$$E\left[\frac{Z^i(a_{t+1})}{\langle x^*, Z(a_{t+1})\rangle} \mid \omega^t\right] = E\frac{Z^i(a_{t+1})}{\langle x^*, Z(a_{t+1})\rangle} = 1$$

by virtue of the necessary and sufficient conditions for the log-optimality of x^*. This completes the proof. □

A Model with i.i.d. Prices We are going to consider a version of the previous model that will make it possible to demonstrate a paradoxical phenomenon related to capital growth. A characteristic feature of this model is that the asset prices S_t, rather than asset returns R_t (or gross returns $Z_t = 1 + R_t$) are i.i.d. As before, the states $a_0, a_1, \ldots$ realized at times $t = 0, 1, \ldots$ are assumed to be i.i.d., and the probability distribution of each a_t is specified by L numbers $p(a^1) > 0$, $p(a^2) > 0, \ldots, p(a^L) > 0$ whose sum is equal to one. Here it will be convenient to assume that the process $a_0, a_1, \ldots$ starts at time $t = 0$, rather than $t = 1$.

There are two assets (e.g., currencies) $i = 0, 1$. The price S_t^0 of asset $i = 0$ is always equal to 1, the price of asset $i = 1$ (the exchange rate) is a given function $S^1(a)$ of the state a. At time $t = 0, 1, \ldots$, the price is $S_t^1 = S^1(a_t)$. Thus the price vector is $(1, S^1(a_t))$.

As long as the price vector is

$$(1, S^1(a_t)),$$

the vector of (gross) returns

$$Z_t^i = S_t^i / S_{t-1}^i, \; i = 0, 1,$$

is given by

$$Z_t = (Z_t^0, Z_t^1) = (1, S^1(a_t)/S^1(a_{t-1})).$$

In what follows, we will write $S(a_t)$ in place of $S^1(a_t)$ to alleviate the notation. Thus the vector Z_t can be written

$$Z_t = (1, S(a_t)/S(a_{t-1})).$$

Trading Strategies and Wealth Dynamics Recall that a trading strategy $H = (h_0, h_1, \ldots)$ with investment proportions $x_0, x_1, \ldots$ (x_t being a vector in the unit simplex) is called *simple*, if x_t does not depend on t and on the observed history states of the world $\omega^t = (a_1, \ldots, a_t)$. A strategy like that is defined by the trader's initial wealth $w > 0$ and a vector $x = (x^0, x^1)$ such that $x^0 \geq 0$, $x^1 \geq 0$ and $x^0 + x^1 = 1$. The numbers x^0 and x^1 indicate the investment proportions for assets $i = 0$ and $i = 1$. The following formula was derived for the trader's wealth (portfolio value) V_t^H at time t:

$$V_t^H = w \langle x, Z_1 \rangle \ldots \langle x, Z_t \rangle, \tag{18.4}$$

where in the present model,

$$\langle x, Z_t \rangle = x^0 Z_t^0 + x^1 Z_t^1 = x^0 + x^1 S(a_t)/S(a_{t-1}). \tag{18.5}$$

18.4 Volatility-Induced Growth

A Puzzle Consider the following question. Assume that the asset price vectors are i.i.d. (this is so in the model under consideration). Let $H = (h_0, h_1, \ldots)$ be a simple trading strategy with some positive initial wealth w and fixed vector of strictly positive investment proportions x. Suppose a trader uses this strategy during a sufficiently long time period. What will happen with the trader's wealth V_t^H in the long run (as time $t \to \infty$)?

Will it tend to:

(a) Fluctuate randomly, converging in one sense or another to a stationary process?
(b) Increase?
(c) Decrease?

Common Intuition Common intuition suggests that if the market is stationary, then the portfolio value V_t^H for a constant proportions strategy must converge in one sense or another to a stationary process. If the correct answer is not known in advance, a typical choice among the above three options is (a). The usual intuitive argument in support of this conjecture appeals to the self-financing property of a constant proportions strategy. The self-financing constraint seems to exclude possibilities of unbounded growth. This argument is also substantiated by the fact that in the deterministic case both the prices and the portfolio value are constant. Indeed, if the price S_t of the risky asset does not depend on t, then $Z_t^1 = S_t/S_{t-1} = 1$, and the vector of gross returns is $Z_t = (1, 1)$. Its scalar product with any vector of proportions x is equal to 1:

$$\langle x, Z_t \rangle = x^0 \cdot 1 + x^1 \cdot 1 = 1,$$

and the formula for the portfolio value gives

$$V_t^H = w \langle x, Z_1 \rangle \ldots \langle x, Z_t \rangle = w \cdot 1 \cdot 1 \cdot \ldots \cdot 1 = w.$$

This way of reasoning makes the answer (a) to the above question seemingly more plausible than the others.

Volatility-Induced Growth Paradoxically, the correct answer to the above question is (b). Under general assumptions, the trader's wealth will not only tend to increase, but it will converge to infinity at an exponential rate with probability one. The precise statement of the result is as follows.

Theorem 18.3 *Let the following conditions hold:*

(i) There are at least two states of the world a and a' for which the price of the risky asset takes on two different values:

$$S(a) \neq S(a').$$

(ii) The given vector of proportions $x = (x^0, x^1)$ has strictly positive coordinates.

Then the wealth V_t^H of the investor using the simple strategy with the vector of proportions x and initial wealth $w > 0$ tends to infinity at an exponential rate with probability one.

Assumption (i) says that there is some randomness (volatility) in the price process of the risky asset. We have seen that if this assumption is not valid, then the market is essentially deterministic, and the result fails to hold. Thus, volatility (which is usually regarded as an impediment to financial growth) may be regarded here as an "engine" of it. In this connection, the phenomenon described in Theorem 18.3 is referred to as *volatility induced growth*.

Proof of Theorem 18.3

1st step. We first observe that in order to prove Theorem 18.3 it is sufficient to show that $E \ln\langle x, Z_t\rangle > 0$. Note that this expectation does not depend on t because for $x = (x^0, x^1)$, we have

$$\langle x, Z_t\rangle = x^0 + x^1 \frac{S(a_t)}{S(a_{t-1})}, \tag{18.6}$$

and so $\ln\langle x, Z_t\rangle = f(a_{t-1}, a_t)$, where

$$f(a_{t-1}, a_t) = \ln\left[x^0 + x^1 \frac{S(a_t)}{S(a_{t-1})}\right].$$

Since the states $a_0, a_1, a_2, \ldots$ are i.i.d., the expectation $Ef(a_{t-1}, a_t)$ does not depend on t.
We know that $V_t^H = w\langle x, Z_1\rangle \ldots \langle x, Z_t\rangle$. Consequently, with probability one,

$$\lim_{t\to\infty} \frac{1}{t} \ln V_t^H$$

$$= \lim\left[\frac{1}{t}\ln w + \frac{\ln\langle x, Z_1\rangle + \ldots + \ln\langle x, Z_t\rangle}{t}\right]$$

$$= \lim \frac{f(a_0, a_1) + \ldots + f(a_{t-1}, a_t)}{t} = Ef(a_{t-1}, a_t) = E\ln\langle x, Z_t\rangle$$

by virtue of the Law of Large Numbers.

We use here the following version of the Law of Large Numbers:
If $a_0, a_1, \ldots$ are i.i.d., then for a function $f(a_{t-1}, a_t)$ the averages

$$\frac{1}{t}(f(a_0, a_1) + \ldots + f(a_{t-1}, a_t))$$

converge to the expectation $Ef(a_{t-1}, a_t)$ with probability one.

If the number $\kappa = E \ln\langle x, Z_t\rangle$ is strictly positive and the sequence

$$\kappa_t = \frac{1}{t} \ln V_t^H, \; t = 1, 2, \ldots,$$

converges to this number, then all elements of this sequence will be greater than $\kappa/2 > 0$ for all t large enough. Thus

$$\frac{1}{t} \ln V_t^H > \frac{\kappa}{2}$$

and so

$$V_t^H > e^{t(\kappa/2)}$$

for all t large enough. This means that $V_t^H \to \infty$ (at an exponential rate!) with probability one.

2nd step. It remains to show that

$$E \ln\langle x, Z_t\rangle = E \ln\left[x^0 + x^1 \frac{S(a_t)}{S(a_{t-1})}\right] > 0.$$

Since the states a_{t-1} and a_t are i.i.d., and the probability that $a_t = a$ is equal to $p(a)$, the above expectation can be written as

$$E \ln\left[x^0 + x^1 \frac{S(a_t)}{S(a_{t-1})}\right] = \sum_{a,a'} p(a)p(a') \ln\left[x^0 + x^1 \frac{S(a')}{S(a)}\right],$$

where a and a' range through the state space A.

Thus we need to prove that the sum

$$\sum_{a,a'} p(a)p(a') \ln\left[x^0 + x^1 \frac{S(a')}{S(a)}\right] \tag{18.7}$$

is strictly positive. To this end we will use the fact that the function $g(y) = \ln y$ is strictly concave.

Recall the definition: a function $g(y)$ is *strictly concave* if for any two distinct points u and v in its domain and for each pair of numbers $x^0 > 0$ and $x^1 > 0$ with $x^0 + x^1 = 1$, we have:

$$g(x^0 u + x^1 v) > x^0 g(u) + x^1 g(v).$$

This is so if *the second derivative g'' is negative*. Clearly, if $u = v$, then the above inequality turns into equality, and so the non-strict inequality holds for any pair u, v.

By virtue of the strict concavity of the logarithm, we get

$$\ln\left[x^0 + x^1 \frac{S(a')}{S(a)}\right] \geq x^0 \ln 1 + x^1 \ln \frac{S(a')}{S(a)} = x^1 \ln \frac{S(a')}{S(a)} \tag{18.8}$$

with a strict inequality when $S(a')/S(a) \neq 1$. We have assumed that $S(a') \neq S(a)$ for some a and a', and so for some pair a, a' the inequality in (18.8) is strict. Therefore the sum in (18.7) is strictly greater than the sum

$$\sum_{a,a'} p(a)p(a')x^1 \ln \frac{S(a')}{S(a)}$$

But this sum is equal to zero because

$$\sum_{a,a'} p(a)p(a') \ln \frac{S(a')}{S(a)}$$

$$= \sum_{a,a'} p(a)p(a')[\ln S(a') - \ln S(a)]$$

$$= \sum_{a,a'} p(a)p(a') \ln S(a') - \sum_{a,a'} p(a)p(a') \ln S(a)$$

$$= \sum_{a'} p(a') \ln S(a') - \sum_{a} p(a) \ln S(a) = 0$$

(we used here the fact that $\sum p(a) = \sum p(a') = 1$).

This completes the proof. □

General Equilibrium Analysis of Financial Markets

19

19.1 Walrasian Equilibrium

The Walrasian Equilibrium Concept In the previous chapters, we presented a theory of the pricing of derivative securities. This theory assumes that the prices of basic (underlying) assets have already been established. But how are the prices of basic assets formed? An answer to this question is provided by *equilibrium models of financial markets*.

The most common, well-examined equilibrium models are those relying upon the ideas of Léon Walras—one of the key figures in the economic thought of the 19th century. According to the Walrasian paradigm, market participants aim to maximize their utilities of consumption subject to budget constraints expressed in terms of the prevailing prices. This generates individual demand (for commodities or assets) depending on the current price system. In equilibrium, aggregate market demand is equal to aggregate supply, which determines the equilibrium prices.

The first rigorous and general mathematical model in which the Walrasian ideas were implemented was proposed in 1950s by Arrow and Debreu[1]—the founders of *General Equilibrium Theory.* A version of the Arrow-Debreu framework aimed at the modelling of financial markets was developed by Radner in 1970s. In this chapter, we review Radner's model and present some basic facts related to it.

The Data of the Model

- There are two moments of time (dates) 0 and 1.
- There is a finite set $A = \{1, 2, \dots, L\}$ of L states of the world (labelled by the natural numbers $a = 1, 2, \dots, L$). The state of the world $a \in A$ is realized at time 1. At time 0, its realization is not known.

[1]Nobel Laureates in Economics (1972 and 1983, respectively).

I.V. Evstigneev et al., *Mathematical Financial Economics*, Springer Texts in Business and Economics, DOI 10.1007/978-3-319-16571-4_19

- There are N securities (assets) $i = 1, 2, \ldots, N$ traded on the market at time 0. Each security i is characterized by its *payoff function* $V^i(a)$. The payoff is obtained at time 1, and it depends on the state of the world a at that time. A security may be regarded as a *contingent contract* whose owner can receive the amount $V^i(a)$ depending on the state of the world a. We denote by

$$V(a) = (V^1(a), \ldots, V^N(a))$$

  the vector of payoffs of the N securities.
- There are K *economic agents* (investors, traders) $k = 1, 2, \ldots, K$ acting in the market. They buy and sell securities at time 0. An agent k, who has selected a *portfolio*

$$h_k = (h_k^1, \ldots, h_k^N)$$

  (of "physical units") of assets at time 0, receives the amount

$$\langle h_k, V(a) \rangle = h_k^1 V^1(a) + \ldots + h_k^N V^N(a)$$

  at time 1. Portfolio positions h_k^i can be both positive and negative.
- Each agent k has an *endowment* (an amount of money) $w_k(0)$ at time 0 and $w_k(a)$, $a = 1, \ldots, L$, at time 1. The endowment at time 1 depends on the state of the world at that time. The vector

$$w_k = (w_k(0), w_k(1), \ldots, w_k(L))$$

  describes the *income stream* which agent k receives exogenously. The vector w_k has dimension $L + 1$ (the number of the states of the world plus one).
- *Preferences* of agent k are characterized by the *utility function*

$$U_k(x) = U_k(x(0), x(1), \ldots, x(L))$$

  defined on the set of all income streams, regarded here as *consumption plans* ($x(0)$ is consumed at time 0; $x(a)$ is consumed at time 1 at state a). Suppose there are two consumption plans $x = (x(0), x(1), \ldots, x(L))$ and $y = (y(0), y(1), \ldots, y(L))$. Then agent k would prefer x to y if $U_k(x) > U_k(y)$ and would be indifferent between x and y if $U_k(x) = U_k(y)$.

This is all what is given in this model. What is not given are the *asset prices* at time 0, when the assets are traded. The purpose of the model is to determine these prices based on the concept of equilibrium. Two basic conditions define equilibrium:

(a) all agents maximize their preferences;
(b) markets for all assets clear.

Agent k's Optimization Problem Suppose prices $\pi^1, \dots, \pi^N$ on assets $i = 1, 2, \dots, N$ are fixed. If agent k constructs a portfolio $h = (h^1, \dots, h^N)$ at time 0, then he/she has to pay for it

$$\langle \pi, h \rangle = \pi^1 h^1 + \dots + \pi^N h^N, \text{ where } \pi = (\pi^1, \dots, \pi^N),$$

and at time 1 he/she gets

$$\langle V(a), h \rangle = V^1(a)h^1 + \dots + V^N(a)h^N,$$

(depending on a). The amount $\langle \pi, h \rangle$ is subtracted from the agent k's endowment $w_k(0)$ available at time 0, and the amount $\langle V(a), h \rangle$ is added to the endowment $w_k(a)$ available at time 1 at state a. This leads to the following consumption plan

$$(w_k(0) - \langle \pi, h \rangle,\ w_k(1) + \langle V(1), h \rangle, \dots, w_k(L) + \langle V(L), h \rangle). \tag{19.1}$$

Denote this consumption plan by $c_k(\pi, h)$. It depends on the asset price vector π, on the portfolio h and on the endowments of agent k.

Each agent k maximizes the utility of consumption, solving the following optimization problem:

($\mathbf{M}_k(\pi)$) *Given the price vector π, select that portfolio h which maximizes the utility $U_k(c_k(\pi, h))$ of the consumption plan $c_k(\pi, h)$ subject to the constraint $c_k(\pi, h) \geq 0$.*

The inequality $c_k(\pi, h) \geq 0$ in the optimization problem means that all the coordinates of the vector $c_k(\pi, h)$ are nonnegative (the amounts consumed are nonnegative).

Radner Equilibrium The model described is a version of that proposed by Radner (1972). The central concept in it is as follows.

Definition Let $\pi \geq 0$ be a vector of asset prices and let

$$h_1, \dots, h_K$$

be portfolios of assets of K agents $k = 1, 2, \dots, K$. We say that $(\pi, h_1, \dots, h_K)$ is an *equilibrium* if the following conditions hold:

(i) For each $k = 1, 2, \dots, K$, the portfolio h_k solves the optimization problem ($\mathbf{M}_k(\pi)$);
(ii) The market for each asset i clears:

$$h_1^i + \dots + h_K^i = 0,\ i = 1, 2, \dots, N.$$

According to (i), every investor selects the most preferred portfolio (given the equilibrium prices π). Condition (ii) expresses the fact that the market for each asset is in zero net supply: every asset (contract) purchased by any agent is issued (written) by some other agent.

19.2 On the Existence of Equilibrium

Theorem 19.1 *An equilibrium exists under the following assumptions:*

(a) *The utility functions of all the agents are strictly increasing (in each coordinate) and concave.*
(b) *All the endowments are strictly positive.*
(c) *There exists a portfolio with positive non-zero payoff.*

The proof requires advanced mathematical techniques and it will be omitted. The reader is referred to the textbook by Magill and Quinzii (2002, Theorem 10.5) (see the list of references at the end of this book).

Assumptions (a)–(c) will be supposed to hold throughout the remainder of this chapter.

Real and Financial Balance Consider the equilibrium consumption plan

$$c_k(\pi, h_k) = (c_k(0), c_k(1), \ldots, c_k(L))$$

for agent k corresponding to the equilibrium price vector π and the equilibrium portfolio h_k. The following balance relation holds:

$$\sum_{k=1}^{K} c_k(l) = \sum_{k=1}^{K} w_k(l),\ l = 0, 1, \ldots, L. \tag{19.2}$$

For $l = 0$, this relation means that at time 0 the total amount $c(0) = \sum c_k(0)$ consumed by all the agents is equal to their aggregate initial endowment $w(0) = \sum w_k(0)$. For $l = 1, \ldots, L$, the meaning of relation (19.2) is similar: at time 1 at each state of the world $a = l$, the total amount consumed, $c(l) = \sum c_k(l)$, is equal to the aggregate endowment, $w(l) = \sum w_k(l)$.

To prove (19.2) we use the market clearing condition

$$h_1^i + \ldots + h_K^i = 0,\ i = 1, 2, \ldots, N. \tag{19.3}$$

To establish (19.2) for $l = 0$ we write

$$\sum_{k=1}^{K} c_k(0) = \sum_{k=1}^{K} [w_k(0) - \langle \pi, h_k \rangle] = \sum_{k=1}^{K} w_k(0),$$

where the second equality holds because

$$\sum_{k=1}^{K} \langle \pi, h_k \rangle = \sum_{k=1}^{K} \sum_{i=1}^{N} \pi^i h_k^i = \sum_{i=1}^{N} \pi^i \sum_{k=1}^{K} h_k^i = 0$$

[see (19.3)]. The proof of (19.2) for $l = 1, \ldots, L$ is analogous (replace in the last two formulas $-\langle \pi, h_k \rangle$ by $\langle V(l), h_k \rangle$ and π^i by $V^i(l)$).

Agents' Beliefs Suppose that each agent $k = 1, 2, \ldots, K$ has his/her own beliefs about the probabilities of states $a = 1, 2, \ldots, L$ at time 1. Assume that for agent k these probabilities are

$$p_k(1) > 0, \ldots, p_k(L) > 0$$

We will denote this probability measure on the set $A = \{1, 2, \ldots, L\}$ by P_k and the expectation with respect to it by E^{P_k}. Thus for any random variable $\xi(a)$ (depending in the random state of the world a), we have

$$E^{P_k} \xi = \sum_{a=1}^{L} p_k(a) \xi(a).$$

Von Neumann-Morgenstern Utilities We will assume that agent k's utility function $U_k(x)$ has the following structure. If $x = (x(0), x(1), \ldots, x(L))$ is a consumption plan, then

$$U_k(x) = u_{k0}(x(0)) + \sum_{a=1}^{L} p_k(a) u_{k1}(x(a)), \tag{19.4}$$

where the function u_{k0} represents the utility of money for agent k at time 0 and u_{k1} the utility of money for agent k at time 1. If the utility function $U_k(x)$ can be written in the form (19.4)—involving the expectation with respect to agent k's subjective probability P_k—then $U_k(x)$ is said to have a *von Neumann-Morgenstern representation*.

In line with assumption (a) of Theorem 19.1, we assume that u_{k0} and u_{k1} are *strictly increasing, concave and twice continuously differentiable*.

19.3 Rational Expectations and Equilibrium Pricing

Equilibrium Pricing Formula Let $\pi \geq 0$ be a vector of equilibrium asset prices and let $h_1, \ldots, h_K$ be the portfolios of agents they select under these prices, maximizing their utilities. Consider the equilibrium consumption plan $c_k(\pi, h_k)$ of agent k:

$$(c_k(0), c_k(1), \ldots, c_k(L)) \tag{19.5}$$
$$= (w_k(0) - \langle \pi, h_k \rangle, w_k(1) + \langle V(1), h_k \rangle, \ldots, w_k(L) + \langle V(L), h_k \rangle).$$

To simplify technical matters, we will assume that all the amounts $c_k(0), c_k(1), \ldots,$ $c_k(L)$ consumed at times 0 and 1 are strictly positive.

Suppose that agent k changes the ith position of his/her portfolio $h_k = (h_k^1, \ldots, h_k^N)$ by adding γ units of asset i (or subtracting if γ is negative). This will lead to the following consumption plan

$$(c_k(0) - \pi^i \gamma, c_k(1) + V^i(1)\gamma, \ldots, c_k(L) + V^i(L)\gamma). \tag{19.6}$$

Its expected utility is

$$\phi(\gamma) = u_{k0}(c_k(0) - \pi^i \gamma) + \sum_{a=1}^{L} p_k(a) u_{k1}(c_k(a) + V^i(a)\gamma).$$

For all γ in a sufficiently small interval $(-\varepsilon, \varepsilon)$ $(\varepsilon > 0)$ the consumption plan (19.6) is feasible (non-negative), and its expected utility is not greater than the expected utility of the equilibrium consumption plan (19.5). Thus $\phi(\gamma)$ attains its maximum at $\gamma = 0$, and so $\phi'(0) = 0$. By differentiating $\phi(\gamma)$ and setting $\gamma = 0$, we obtain

$$-\pi^i u'_{k0}(c_k(0)) + \sum_{a=1}^{L} p_k(a) V^i(a) u'_{k1}(c_k(a)) = 0.$$

This relation leads to the following *equilibrium pricing formula*:

$$\pi^i = \sum_{a=1}^{L} p_k(a) V^i(a) \frac{u'_{k1}(c_k(a))}{u'_{k0}(c_k(0))}. \tag{19.7}$$

This is a central result of the equilibrium theory.

Marginal Rates of Substitution The numbers

$$W_k(a) = \frac{u'_{k1}(c_k(a))}{u'_{k0}(c_k(0))} \tag{19.8}$$

are called the *marginal intertemporal rates of substitution*. They are computed by comparing the marginal utilities of the amounts $c_k(0)$ at time 0 and $c_k(a)$ at time 1. The function $W_k(a)$ of the state a—regarded as a random variable—is called the *pricing kernel*. Note that $W_k(a) > 0$ because $u'_{k0} > 0$ and $u'_{k1} > 0$ [see (19.8)]. We can write formula (19.7) in terms of the pricing kernel $W_k(a)$ and the expectation E^{P_k} with respect to the subjective probability P_k of agent k:

$$\pi^i = E^{P_k}[W_k(a)V^i(a)]. \tag{19.9}$$

Remark 19.1 The left-hand side of (19.9) does not depend on k. This means that we can compute the equilibrium price π^i of asset i in terms of the subjective probabilities P_k and other attributes of *any* agent k. The result will be independent of k.

Rational Expectations In what follows, we will assume that the agents have *homogeneous and correct beliefs* (also referred to as *rational expectations*), i.e. all the subjective probabilities P_k coincide with the "real world" probability P. Then the equilibrium pricing formula becomes

$$\pi^i = E[W_k(a)V^i(a)], \tag{19.10}$$

where E is the expectation with respect to P.

Equilibrium Interest Rate Suppose now there are $N + 1$, rather than N, assets in the model, $i = 0, 1, \dots, N$, and the payoff of the 0th, $V^0(a) > 0$, is the same for all states a: $V^0(a) = V^0$. This means that the asset is *riskless*.

Let us fix some k and write $W(a)$ in place of $W_k(a)$. Then the pricing formula (19.10) applied to the riskless asset gives:

$$\pi^0 = V^0 EW(a). \tag{19.11}$$

Define the number r so that $1 + r = V^0/\pi_0$. Then $V^0 = (1 + r)\pi^0$, which means that r is the *equilibrium interest rate*—the rate of return on the riskless security.

Let us write EW in place of $EW(a)$. Formula (19.11) makes it possible to express the equilibrium interest rate r through the expectation of the pricing kernel:

$$1 + r = 1/EW. \tag{19.12}$$

Typically, in equilibrium, $r > 0$ (under normal conditions because agents discount their utilities u_{k1} when evaluating future income). In what follows, we assume that $r > 0$.

Dynamic Securities Market Model with Equilibrium Prices Consider a dynamic securities market model (studied in detail in the previous chapters) in which the prices are obtained from the Radner equilibrium model. Assume that there are two moments of time $t = 0, 1$ (i.e., $T = 1$), the state space is $A = \{1, 2, \dots, L\}$, and the prices of the $N + 1$ assets are as follows. Define the prices of the N risky assets by

$$S_0^i = \pi^i, \; S_1^i(a) = V^i(a), \; i = 1, 2, \dots, N, \tag{19.13}$$

and the prices of the riskless asset are defined by

$$S_0^0 = 1, \; S_1^0 = 1 + r. \tag{19.14}$$

Here $V^i(a)$ are the payoffs of the assets at time 1 given in the equilibrium model, π^i are the equilibrium prices and r is the equilibrium interest rate. The price vectors at times 0 and 1 can be written as

$$\mathbf{S}_0 = (1, S_0^1, \dots, S_0^N), \; \mathbf{S}_1(a) = (1 + r, S_1^1(a), \dots, S_1^N(a)). \tag{19.15}$$

19.4 Arbitrage and Equilibrium

Absence of Arbitrage A central result connecting the two pricing approaches (equilibrium and no-arbitrage), as well as the two models, is as follows.

Proposition 19.1 *Equilibrium prices do not allow for arbitrage opportunities.*

Thus, if the price vectors at times 0 and 1 and the interest rate r are defined by (19.13)–(19.15), the one-period dynamic securities market model defined above is arbitrage-free.

Risk-Neutral Probability We know that the no-arbitrage hypothesis holds if and only if there exists a risk neutral probability measure. We will construct such a measure, and thus prove Proposition 19.1. Fix some $k = 1, 2, \dots, K$ and consider the pricing kernel $W_k(a)$. Let us write $W(a)$ in place of $W_k(a)$ for shortness. According to the formulas for the equilibrium prices and the equilibrium interest rate r [see (19.10) and (19.12)], we have

$$\pi^i = E[W(a)V^i(a)], \; i = 1, 2, \dots, N; \; 1 + r = 1/EW. \tag{19.16}$$

Suppose the given measure P (with respect to which the expectations in (19.16) are taken) assigns the probabilities $p(a)$ to the states $a = 1, 2, \dots, L$. Define a new measure Q for which the probabilities of the states $a = 1, 2, \dots, L$ are given by

$$q(a) = (1 + r)W(a)p(a). \tag{19.17}$$

Proposition 19.2 *Formula (19.17) defines a risk-neutral probability measure Q.*

Proof

1st step. Let us first show that Q is indeed a probability measure. We have $q(a) > 0$ because $W(a) > 0$ and $p(a) > 0$. Furthermore,

$$\sum_{a=1}^{A} q(a) = (1+r)\sum_{a=1}^{A} W(a)p(a) = (1+r)EW = 1$$

[see (19.16)].

2nd step. Denote by E^Q the expectation with respect to Q. Let us show that

$$\mathbf{S}_0 = \frac{E^Q\mathbf{S}_1(a)}{1+r}. \qquad (19.18)$$

This equality should be verified for each of the $N+1$ coordinates of the vectors $\mathbf{S}_0$ and $\mathbf{S}_1(a)$ [see (19.15)]. For the 0th coordinate, the left-hand side is 1, and on the right-hand side, we have $E^Q(1+r)/(1+r) = 1$. For the ith coordinate ($i = 1, 2, \ldots, N$), we have

$$(1+r)^{-1}E^Q S_1^i(a) = (1+r)^{-1}\sum_{a=1}^{A}(1+r)W(a)p(a)S_1^i(a)$$

$$= \sum_{a=1}^{A} W(a)p(a)S_1^i(a) = E(WS_1^i) = E(WV^i) = \pi^i = S_0^i$$

by virtue of (19.13) and (19.16).

3rd step. It remains to show that the net present value of any self-financing strategy $(\mathbf{h}_0, \mathbf{h}_1)$ has zero expectation E^Q, i.e.

$$E^Q[\frac{\langle \mathbf{S}_1, \mathbf{h}_1\rangle}{1+r} - \langle \mathbf{S}_0, \mathbf{h}_0\rangle] = 0.$$

By virtue of self-financing, $\langle \mathbf{S}_1, \mathbf{h}_1\rangle = \langle \mathbf{S}_1, \mathbf{h}_0\rangle$, and so

$$E^Q\frac{\langle \mathbf{S}_1, \mathbf{h}_1\rangle}{1+r} = E^Q\frac{\langle \mathbf{S}_1, \mathbf{h}_0\rangle}{1+r} = \frac{\langle E^Q\mathbf{S}_1, \mathbf{h}_0\rangle}{1+r} = \langle \mathbf{S}_0, \mathbf{h}_0\rangle$$

in view of (19.18) and because $\mathbf{h}_0$ is non-random.

The proof is complete. □

Remark 19.2 The last result not only proves the no-arbitrage hypothesis, but also demonstrates an important link between the notions of a risk-neutral probability measure and a pricing kernel.

20 Behavioral Equilibrium and Evolutionary Dynamics

20.1 A Behavioral Evolutionary Perspective

Market Equilibrium and Dynamics: A New Perspective This chapter reviews a new class of models that have been developed in the last decade as an attempt to find a plausible alternative to Walrasian General Equilibrium. The classical models (such as that considered in the previous chapter) assume that market participants are fully rational and act so as to maximize utilities of consumption subject to budget constraints. The models we are going to consider now are based on a different paradigm of *behavioral equilibrium*, admitting that market participants are boundedly rational and may have a whole variety of patterns of behavior determined by their individual psychology (not necessarily describable in terms of individual utility maximization).

A Synthesis of Behavioral and Evolutionary Approaches Patterns of strategic behavior of investors/traders may involve, for example, mimicking, satisficing, rules of thumb based on experience, etc. They might be *interactive:* depending on the behavior of the others, and *relative*: taking into account the comparative performance of the others. Objectives of the market participants might be of an *evolutionary* nature: *survival* (especially in crisis environments), *domination* in a market segment, fastest capital *growth*, etc. The evolutionary aspect is in the main focus of the model described below. A synthesis of behavioral and evolutionary approaches makes it possible to identify strategies guaranteeing survival or domination in a competitive market environment.

I.V. Evstigneev et al., *Mathematical Financial Economics*, Springer Texts in Business and Economics, DOI 10.1007/978-3-319-16571-4_20

The roots of ideas underlying these models lie in *Evolutionary Economics* (Alchian 1950s, Nelson and Winter 1970s—1990s), *Behavioral Economics* (Tversky, Kahneman and Smith[1]), and *Behavioral Finance* (Shiller[2]).

Model Description: The Basic Elements

- There is a stochastic process $a_1, a_2, \dots$ with values in a finite set A, where $a_t \in A$ ($t = 1, 2, \dots$) is the "state of the world" at date t. To simplify presentation we will assume that a_t are independent and identically distributed (i.i.d.).
- N *securities (assets)* $n = 1, 2, \dots, N$ are traded on the market at each of the dates $t = 0, 1, \dots$. At each date t, assets pay *dividends* $D_{t,n} = D_n(a_t) \geq 0$ depending on the state of the world a_t at date t. The total mass (the number of "physical units") of asset n available at each date t is $V_{t,n} = V_{t,n}(a_t) > 0$.
- K *investors/traders* $k = 1, \dots, K$ can buy and sell assets at each of the dates $t = 0, 1, \dots$ Investor k's *portfolio* at date $t = 0, 1, 2, \dots$ is

$$x_t^k = (x_{t,1}^k, \dots, x_{t,N}^k),$$

  where $x_{t,n}^k$ is the holding (the number of units) of asset n. The portfolio $x_t^i = x_t^i(\omega^t)$ for $t \geq 1$ is selected based on the observation of the history of states of the world up to date t

$$\omega^t = (a_1, \dots, a_t).$$

  It will be assumed here (to alleviate presentation) that the coordinates of the portfolio vectors are non-negative: short selling will be ruled out.
- We denote by $p_t = (p_{t,1}, \dots, p_{t,N})$ the vector of equilibrium *asset prices* at date t, where $p_{t,n} \geq 0$ stands for the price of one unit of asset n at date t. The prices depend on the history $\omega^t = (a_1, \dots, a_t)$ of states of the world:

$$p_{t,n} = p_{t,n}(\omega^t).$$

  The market value of a portfolio x_t^k at date t:

$$\langle p_t, x_t^k \rangle := \sum_{n=1}^{N} p_{t,n} x_{t,n}^k.$$

[1]Kahneman and Smith: the 2002 Nobel Laureates in Economics.

[2]The 2013 Nobel Prize in Economics.

Investors' Budgets At date $t = 0$ investors $k = 1, 2, \ldots, K$ have *initial endowments* $w_0^k > 0$. Investor k's *budget* at date $t \geq 1$ is

$$B_t^k(p_t, x_{t-1}^k) := \langle D_t + p_t, x_{t-1}^k \rangle, \tag{20.1}$$

where

$$D_t = D(a_t) = (D_1(a_t), \ldots, D_N(a_t)).$$

The budget consists of two components: the dividends $\langle D_t, x_{t-1}^k \rangle$ paid by the yesterday's portfolio and the current market value $\langle p_t, x_{t-1}^k \rangle$ of this portfolio.

The fraction $\alpha_t = \alpha_t(a_t) \in (0, 1)$ of the budget is invested into assets. Factors that might determine the *investment rate* α_t are tax, transaction costs, and consumption.

Investment Strategies Investor k allocates wealth across the assets according to investment proportions

$$\lambda_t^k = (\lambda_{t,1}^k, \ldots, \lambda_{t,N}^k) \geq 0, \ \sum_n \lambda_{t,n}^k = 1.$$

An *investment strategy* (*portfolio rule*) Λ^k of investor k is a sequence of vector functions

$$\lambda_t^k = \lambda_t^k(\omega^t), \ t = 0, 1, \ldots,$$

specifying the vectors of k's investment proportions depending on the observed history $\omega^t = (a_1, \ldots, a_t)$. The *demand function* of investor k is

$$X_{t,n}^k(p_t, x_{t-1}^k) = \frac{\alpha_t \lambda_{t,n}^k B_t^k(p_t, x_{t-1}^k)}{p_{t,n}}, \ n = 1, \ldots, N.$$

Short-Run (Temporary) Equilibrium is determined by the market clearing condition

$$\sum_{k=1}^K X_{t,n}^k(p_t, x_{t-1}^k) = V_{t,n} \ , \ n = 1, 2, \ldots, N.$$

This system of equations can be written as

$$\sum_{k=1}^K \alpha_t \lambda_{t,n}^k B_t^k(p_t, x_{t-1}^k) = p_{t,n} V_{t,n}, \ n = 1, 2, \ldots, N. \tag{20.2}$$

It can be easily shown that the pricing equation has a unique solution $p_{t,n} \geq 0$.

We will consider only *admissible* strategy profiles—those for which $p_{t,n} > 0$: only in this case the above formula for $X^k_{t,n}$ makes sense. It can be shown that if at least one of the portfolio rules λ^k_t has strictly positive investment proportions, then the strategy profile is admissible. This will always be the case in what follows.

Simplifying Assumptions In this introductory presentation, we will impose the following assumptions under which the main results take on a more compact and transparent form:

(i) the investment rate $\alpha_t = \alpha \in (0, 1)$ is constant;
(ii) total volumes $V_{t,n}$ of all assets do not depend on random factors and grow (or decrease) at the same rate $\gamma > \alpha$: $V_{t,n} = \gamma^t V_n$.

It is not excluded of course that $V_{t,n}$ is constant, i.e. $\gamma = 1$.

Equilibrium Market Dynamics Suppose that all the investors $k = 1, 2, \ldots, K$ selected their portfolio rules

$$\Lambda^k = (\lambda^k_0, \lambda^k_1(\omega^1), \lambda^k_2(\omega^2), \ldots).$$

Assume (as we will always assume it in what follows) that the strategy profile $(\Lambda^1, \ldots, \Lambda^K)$ is admissible. Then the equilibrium market dynamics are described by the equations:

$$p_{t,n} V_{t,n} = \sum_{k=1}^{K} \alpha \lambda^k_{t,n} \langle D_t(a_t) + p_t, x^k_{t-1} \rangle, \; n = 1, \ldots, N, \tag{20.3}$$

$$x^k_{t,n} = \frac{\alpha \lambda^k_{t,n} \langle D_t(a_t) + p_t, x^k_{t-1} \rangle}{p_{t,n}}, \; n = 1, \ldots, N, \; k = 1, 2, \ldots, K. \tag{20.4}$$

Equations (20.3) determine the equilibrium prices $p_{t,n}, 1, \ldots, n$, at date t given the investors' portfolios $x^k_{t-1}, k = 1, \ldots, K$, at date t. Formulas (20.4) express investors' portfolios $x^k_t = (x^k_{t,1}, \ldots, x^k_{t,N})$ at date t.

Dynamics of Market Shares We are primarily interested in the comparative performance of investors' strategies characterized in terms of their *relative wealth* (*market shares*)

$$r^k_t = \frac{w^k_t}{w^1_t + \ldots + w^K_t}, \; k = 1, \ldots, K,$$

where $w_t^k := \langle D_t + p_t, x_{t-1}^k \rangle$ is *investor k's wealth*. The dynamics of the vectors $r_t = (r_t^1, \ldots, r_t^K)$ are described by the random dynamical system

$$r_{t+1}^k = \sum_{n=1}^{N} [\rho \langle \lambda_{t+1,n}, r_{t+1} \rangle + (1-\rho) R_{t+1,n}] \frac{\lambda_{t,n}^k r_t^k}{\langle \lambda_{t,n}, r_t \rangle}, \quad k = 1, \ldots, K, \qquad (20.5)$$

where

$$\rho = \alpha / \gamma,$$

$$R_{t,n} = R_n(a_t) = \frac{D_n(a_t) V_n}{\sum_{m=1}^{N} D_m(a_t) V_m}$$

(*relative dividends*) and

$$\langle \lambda_{t+1,n}, r_{t+1} \rangle = \sum_{k=1}^{K} \lambda_{t+1,n}^k r_{t+1}.$$

Given $r_t = (r_t^1, \ldots, r_t^K)$ the vector $r_{t+1} = (r_{t+1}^k, \ldots, r_{t+1}^k)$ is determined as a unique solution to the system of linear equations (20.5), that can be derived from (20.3) and (20.4) via a chain of algebraic transformations.

20.2 Survival Strategies

Definition of Survival Strategies Suppose all the investors $k = 1, 2, \ldots, K$ have selected their strategies $\Lambda^1, \Lambda^2, \ldots, \Lambda^K$. For each k, the strategy profile $(\Lambda^1, \ldots, \Lambda^K)$ generates the random sequence r_t^k, $t = 0, 1, 2, \ldots$ of each investor k's market shares. We say that investor k (or the strategy Λ^k) *survives* in the market selection process investor k's market share is strictly positive and bounded away from zero with probability 1, i.e., there exists a strictly positive number c such that $r_t^k \geq c$ for all t with probability 1.

It is important to note that the property of survival of a strategy Λ^k does not depend solely on Λ^k. It might depend on the whole strategy profile $(\Lambda^1, \ldots, \Lambda^K)$ describing the behavior of the whole group of investors, including the k's rivals $j \neq k$, $j = 1, 2, \ldots, K$. For some strategy profiles, Λ^k might survive, for some not. Therefore the following notion is of importance.

Definition A portfolio rule Λ^k of investor k is called a *survival strategy* if it survives irrespective of what strategies Λ^m all the investors $m \neq k$ use.

The main focus of this theory is on the characterization and analysis of survival strategies.

Strategy Λ^* Denote by $\lambda_1^*, \ldots, \lambda_N^*$ the expectations of the relative dividends of assets $n = 1, \ldots, N$:

$$\lambda_n^* = ER_n(a_t), \text{ where } R_n(a_t) = \frac{D_n(a_t)V_n}{\sum_{m=1}^N D_m(a_t)V_m}$$

and put

$$\lambda^* = (\lambda_1^*, \ldots, \lambda_N^*).$$

Since the states a_t of the world are i.i.d., the expectation $\lambda_n^* = ER_n(a_t)$ does not depend on t.

Consider the strategy

$$\Lambda^* = (\lambda^*, \lambda^*, \lambda^*, \ldots)$$

prescribing to distribute wealth across assets in fixed (constant) proportions $\lambda_1^*, \ldots, \lambda_N^*$, regardless of the moment of time and random situation.

The Main Results Assume that $\lambda^* > 0$.

Theorem 20.1 *The portfolio rule Λ^* is a survival strategy.*

It should be stressed that although Λ^* is a constant proportions strategy, it survives in competition with *all* strategies allowing for any, most general kind of behavior of investors $m \neq k$.

The following result shows that the survival strategy is asymptotically unique.

Theorem 20.2 *If $\Lambda = (\lambda_t)$ is a survival strategy, then*

$$\sum_{t=0}^{\infty} ||\lambda^* - \lambda_t||^2 < \infty \textit{ almost surely.}$$

According to this theorem, the vectors $\lambda_t(\omega^t)$ of investment proportions of any survival strategy $\Lambda = (\lambda_t)$ converge to the vector λ^* with probability 1, and moreover, this convergence is fast enough, so that the sum of the series $\sum_{t=0}^{\infty} ||\lambda^* - \lambda_t||^2$ is finite.

The next theorem expresses the property of *evolutionary stability* of Λ^* in the class of constant proportions strategies. It holds under some additional technical assumptions on the model.

Theorem 20.3 *If among K investors using constant proportions strategies, there is a group following the strategy Λ^*, then those who use Λ^* survive, while all the others are driven out of the market: their market shares tend to zero a.s.*

20.3 Links to the Classical Theory

The Meaning of Λ^* The portfolio rule Λ^* integrates three general principles in Financial Economics.

(a) Λ^* prescribes the allocation of wealth among assets in the proportions of their *fundamental values*—the expectations of the future dividends.
(b) The strategy Λ^*, defined in terms of the relative (weighted) dividends, provides an investment recommendation in line with the CAPM principles, emphasizing the role of the *market portfolio* (see Chap. 7).[3]
(c) The portfolio rule Λ^* is expresses an idea analogous to that of the Kelly portfolio rule of "betting your beliefs": it allocates wealth in the proportions of expected relative dividends (see Chap. 17).

Remark 20.1 We can see that Λ^* does not depend on the investment rate α and the asset growth rate γ. This is so due to our simplifying assumptions, in particular, because a_t are i.i.d. Analogous results hold for more general stochastic processes a_t, but the definition of Λ^* in the general case is more complex.

Comments on the Model

- In the above model, a short-run equilibrium is defined directly in terms of a strategy profile of the agents. A strategy (portfolio rule) specifies an algorithm of investment behavior over each time period $t-1, t$. This algorithm may have quite a general nature, having nothing to do with individual rationality and optimization. The prices in each time period emerge as a result of a *temporary equilibrium of investors' behaviors*. The notion of equilibrium does not involve (typically unobservable) agent's characteristics such as individual utilities and beliefs, and the application of the model does not require the knowledge of these characteristics.
- Conceptually, the model uses the Marshallian "moving equilibrium method" (1920) to describe the dynamics of the asset market as a sequence of consecutive temporary equilibria. To employ this method one needs to distinguish between at least two sets of economic variables changing with *different speeds*. Then the

[3]However, it should be emphasized that instead of weighing assets according to their prices, in Λ^* the weights are based on fundamentals. In practice, Λ^* is an example of *fundamental indexing* (Arnott, Hsu and West, 2008).

set of variables changing *slower* (in our case, the set of vectors of investment proportions) can be temporarily fixed, while the other (in our case, the asset prices) can be assumed to *rapidly* reach the unique state of partial equilibrium.

- In the Walrasian equilibrium model described in the previous chapter, the market participants have to agree on the future prices for each of the possible future realizations of the states of the world (without knowing which particular state will be realized). This assumption—*the perfect foresight hypothesis*—is not needed in the equilibrium framework considered in this chapter. Only historical observations influence current behavior; no agreement about the future events, and no coordination of plans is required.

Problems and Exercises III

21

Question 21.1 (Volatility-Induced Growth Acceleration) Consider a version of the model described in Chap. 18 in which gross returns $Z_t^i = Z^i(a_t) > 0$ of two assets $i = 1, 2$ are functions of the current i.i.d. states of the world a_t, and hence are i.i.d. themselves. (Here, we do not assume that the prices are i.i.d. and that the price of the first asset is equal to 1.) Consider the following analogue of Theorem 18.1.

Let the Following Conditions Hold

(i) *There is at least one state of the world a^* for which the returns on the two assets do not coincide:*

$$Z^1(a^*) \neq Z^2(a^*).$$

(ii) $E \ln Z_t^1 = E \ln Z_t^2 = \gamma > 0.$
Then the wealth V_t^H of the investor using any simple strategy with a strictly positive vector of proportions $x = (x^1, x^2)$ and initial endowment $w > 0$ tends to infinity almost surely at an exponential rate γ' strictly greater than γ.

Interpret the assumptions and the statement of the above result, and prove it.

Answer Assumption (i) says that the *relative return* $Z^1(a)/Z^2(a)$ of the assets exhibits a non-zero volatility (it is not always equal to 1). By virtue of (ii), the price processes $S_t^i = S_0^i Z_1^i \dots Z_t^i$ $(i = 1, 2)$ have the same drift $\gamma > 0$, and hence exhibit exponential growth at the same rate γ with probability one. The above result says that any fully diversified constant proportions strategy leads almost surely to growth of wealth at a rate strictly greater than γ. This phenomenon can be called *volatility induced growth acceleration*. Note that if $Z^1(a) = Z^2(a)$ for all a (the relative

I.V. Evstigneev et al., *Mathematical Financial Economics*, Springer Texts in Business and Economics, DOI 10.1007/978-3-319-16571-4_21

return has zero volatility), then

$$t^{-1} E \ln V_t^H = t^{-1}(\ln w + tE \ln Z_t^1) = E \ln Z_t^1 + t^{-1} \ln w = \gamma + t^{-1} \ln w \to \gamma \text{ (a.s.)}$$

i.e. V_t^H grows at the same exponential rate γ, and no growth acceleration takes place.

To prove the above assertion, we proceed as in the proof of Theorem 18.3. We first show, by using the Law of Large Numbers, that it is sufficient to verify the inequality $E \ln\langle x, Z_t\rangle > \gamma$, where $Z_t = (Z_t^1, Z_t^2) = (Z^1(a_t), Z^2(a_t))$. Then we write

$$E \ln\langle x, Z_t\rangle = E \ln\left[x^1 Z^1(a_t) + x^2 Z^2(a_t)\right] = \sum_{a \in A} p(a) \ln\left[x^1 Z^1(a) + x^2 Z^2(a)\right] >$$
$$\sum_{a \in A} p(a)[x^1 \ln Z^1(a) + x^2 \ln Z^2(a)] = x^1 E \ln Z^1(a_t) + x^2 \ln Z^2(a_t) = \gamma, \tag{21.1}$$

where the last inequality follows from the strict concavity of the function $\ln y$ and the fact that $Z^1(a) \neq Z^2(a)$ for some a.

This completes the proof.

Question 21.2 (Buy Low and Sell High) The above argument yields a rigorous proof of the fact of growth acceleration driven by volatility. But what is the intuition, the underlying fundamental reason for it? There is a simple intuitive explanation of this phenomenon based on the observation that constant proportions always force one to "buy low and sell high"—the common sense dictum of all trading. Those assets whose prices have risen from the last rebalance date will be overweighted in the portfolio, and their holdings must be reduced to meet the required proportions and to be replaced in part by assets whose prices have fallen and whose holdings must therefore be increased. Obviously, for this mechanism to work the prices must change in time; if they are constant, one cannot get any profit from trading. However, such reasoning does not reflect the assumption of *constancy* of investment proportions. This leads to the question: what will happen if the "common sense dictum of all trading" is pushed to the extreme and the portfolio is rebalanced so as to sell *all* those assets that gain value and buy only those ones which lose it? Construct an example showing that in this case, growth acceleration will not necessarily take place.

Answer Assume that there are two assets, the price S_t^1 of the first (riskless) is always 1, and the price S_t^2 of the second (risky) follows a geometric random walk, so that the gross return on it can be either 2 or $1/2$ with equal probabilities. Suppose the investor sells the second asset and invests all wealth in the first if the price S_t^2 goes up and performs the converse operation, betting all wealth on the risky asset, if S_t^2 goes down. Then the sequence $\lambda_t = (\lambda_t^1, \lambda_t^2)$ of the vectors of investment proportions will be i.i.d. with values $(0, 1)$ and $(1, 0)$ taken on with equal probabilities. Furthermore, λ_{t-1} will be independent of R_t. By virtue of (21.1), the growth rate of the portfolio value for this strategy is equal to $E\ln(R_t\lambda_{t-1}) = [\ln(0\cdot 1+1\cdot 2)+\ln(0\cdot 1+1\cdot \frac{1}{2})+\ln(1\cdot 1+0\cdot 2)+\ln(1\cdot 1+0\cdot \frac{1}{2})]/4 = 0$, which is the same as the growth rate of each of the two assets $k = 1, 2$ and is strictly less than the growth rate of any completely mixed constant proportions strategy.

The above example shows that the phenomenon of financial growth driven by volatility depends critically on rebalancing to an arbitrary *fixed* mix of portfolio proportions, and so it may be more delicate than it looks at first glance.

Question 21.3 (Growth Acceleration vs Volatility Reduction) One might conjecture (and there are some hints in this regard in the literature) that an endogenous source of volatility-driven growth might be related to a reduction in volatility ("stabilization") of the wealth process yielded by a constant proportions strategy. In other words, one might suggest that a constant proportions strategy leads to an increase in the growth rate *at the expense of* volatility reduction. To examine whether this conjecture is true let us return to the model analyzed in Question 21.1. Assume that $Z^1 > 0$ and $Z^2 > 0$ are i.i.d. random variables. For any constant proportions strategy

$$(x, 1-x), x \in [0, 1],$$

consider the drift (exponential growth rate) of the wealth process

$$\gamma(x) = E\ln\left[xZ^1 + (1-x)Z^2\right]$$

and its volatility

$$\sigma(x) = \sqrt{Var\{\ln\left[xZ^1 + (1-x)Z^2\right]\}}.$$

It easily seen (verify!) that $\gamma(x)$ is increasing on $[0, 1/2]$. Is it always true that $\sigma(x)$ is decreasing on $[0, 1/2]$?

Answer To construct a counterexample and refute the above conjecture assume that the i.i.d. random variables Z^1 and Z^2 take on values 1 and $a > 0$ with equal probabilities. It is an easy exercise in calculus (left to the reader) to show that there exist some numbers $0 < a_- < 1$ and $a_+ > 1$ with the following properties.

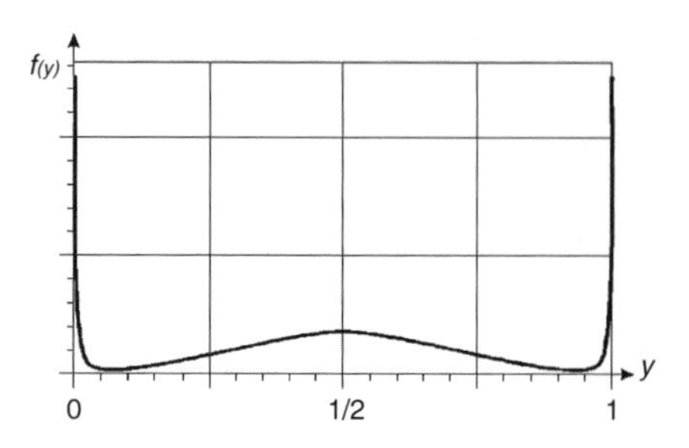

Fig. 21.1 Non-convexity of volatility: an example

(i) If $a \in [a_-, a_+]$, then the function $\sigma(x)$ is decreasing on $[0, 1/2]$ with a minimum at $x = 1/2$.

(ii) If $a \notin [a_-, a_+]$, then it has a local minimum (!) between 0 and $1/2$, and hence it is not monotone decreasing on $[0, 1/2]$. In this case, its graph looks like the one depicted in Fig. 21.1.

The numbers a_- and a_+ are given by

$$a_\pm = 2e^4 - 1 \pm \sqrt{(2e^4-1)^2 - 1},$$

where $a_- \approx 0.0046$ and $a_+ \approx 216.388$.

Question 21.4 (Dynamics of Market Shares) Derive the system of Eq. (20.5) for the dynamics of the market shares r_t^k from Eqs. (20.3) and (20.4).

Answer From (20.3) and (20.4) we get

$$p_{t,n} = V_{t,n}^{-1} \alpha \sum_{k=1}^{K} \lambda_{t,n}^k \langle p_t + D_t, x_{t-1}^k \rangle =$$

$$\alpha V_{t,n}^{-1} \sum_{k=1}^{K} \lambda_{t,n}^k w_t^k = \alpha V_{t,n}^{-1} \langle \lambda_{t,n}, w_t \rangle, \tag{21.2}$$

$$x_{t,n}^k = \frac{V_{t,n} \lambda_{t,n}^k w_t^k}{\langle \lambda_{t,n}, w_t \rangle}, \tag{21.3}$$

where $t \geq 1$, $w_t := (w_t^1, \ldots, w_t^K)$ and $\lambda_{t,n} := (\lambda_{t,n}^1, \ldots, \lambda_{t,n}^K)$. Consequently, we have

$$w_{t+1}^k = \sum_{n=1}^{N} (p_{t+1,n} + D_{t+1,n}) x_{t,n}^k =$$

$$\sum_{n=1}^{N} (\alpha \frac{\langle \lambda_{t+1,n}, w_{t+1} \rangle}{V_{t+1,n}} + D_{t+1,n}) \frac{V_{t,n} \lambda_{t,n}^k w_t^k}{\langle \lambda_{t,n}, w_t \rangle} =$$

$$\sum_{n=1}^{N} (\alpha \frac{\langle \lambda_{t+1,n}, w_{t+1} \rangle V_{t,n}}{V_{t+1,n}} + D_{t+1,n} V_{t,n}) \frac{\lambda_{t,n}^k w_t^k}{\langle \lambda_{t,n}, w_t \rangle}, \ t \geq 0. \tag{21.4}$$

By summing up these equations over $k = 1, \ldots, K$, we obtain

$$W_{t+1} = \sum_{n=1}^{N} (\alpha \frac{\langle \lambda_{t+1,n}, w_{t+1} \rangle V_{t,n}}{V_{t+1,n}} + D_{t+1,n} V_{t,n}) \frac{\sum_{k=1}^{K} \lambda_{t,n}^k w_t^k}{\langle \lambda_{t,n}, w_t \rangle} =$$

$$\sum_{n=1}^{N} (\alpha \frac{\langle \lambda_{t+1,n}, w_{t+1} \rangle V_{t,n}}{V_{t+1,n}} + D_{t+1,n} V_{t,n}).$$

Since

$$V_{t+1,n} / V_{t,n} = \gamma \tag{21.5}$$

does not depend on n , we have

$$W_{t+1} = \sum_{n=1}^{N} (\alpha \gamma^{-1} \langle \lambda_{t+1,n}, w_{t+1} \rangle + D_{t+1,n} V_{t,n}) =$$

$$\sum_{n=1}^{N} (\rho \langle \lambda_{t+1,n}, w_{t+1} \rangle + D_{t+1,n} V_{t,n}) = \rho W_{t+1} + \sum_{n=1}^{N} D_{t+1,n} V_{t,n} \, .$$

This implies the formula

$$W_{t+1} = \frac{1}{1-\rho} \sum_{m=1}^{N} D_{t+1,m} V_{t,m} \, , \tag{21.6}$$

where $\alpha\gamma^{-1} = \rho$. From (21.4) and (21.5), we find

$$w_{t+1}^k = \sum_{n=1}^{N} (\rho\langle\lambda_{t+1,n}, w_{t+1}\rangle + D_{t+1,n} V_{t,n}) \frac{\lambda_{t,n}^k w_t^k}{\langle\lambda_{t,n}, w_t\rangle} \quad t \geq 0.$$

Dividing both sides of this equation by W_{t+1} and using (21.6), we get

$$r_{t+1}^k = \sum_{n=1}^{N} [\rho\langle\lambda_{t+1,n}, r_{t+1}\rangle + (1-\rho)\frac{D_{t+1,n} V_{t,n}}{\sum_{m=1}^{N} D_{t+1,m} V_{t,m}}] \frac{\lambda_{t,n}^k w_t^k / W_t}{\langle\lambda_{t,n}, w_t\rangle / W_t},$$

which yields (20.5).

Question 21.5 (The Case of Two Investors) Show that if there are two investors ($N = 2$), then the dynamics of the ratio $z_t = r_t^1 / r_t^2$ of their market shares is governed by the equation

$$z_{t+1} = z_t \frac{\sum_{n=1}^{N} [\rho\lambda_{t+1,n}^2 + (1-\rho) R_{t+1,n}] \frac{\lambda_{t,n}^1}{\lambda_{t,n}^1 z_t + \lambda_{t,n}^2}}{\sum_{n=1}^{N} [\rho\lambda_{t+1,n}^1 + (1-\rho) R_{t+1,n}] \frac{\lambda_{t,n}^2}{\lambda_{t,n}^1 z_t + \lambda_{t,n}^2}}. \tag{21.7}$$

Answer By using (20.5) with $N = 2$, we get

$$r_{t+1}^k = \sum_{n=1}^{N} [\rho(\lambda_{t+1,n}^k r_{t+1}^k + \lambda_{t+1,n}^j (1 - r_{t+1}^k)) + (1-\rho) R_{t+1,n}] \frac{\lambda_{t,n}^k r_t^k}{\lambda_{t,n}^k r_t^k + \lambda_{t,n}^j r_t^j},$$

where $k, j \in \{1, 2\}$ and $k \neq j$. Setting $C_{t,n}^{kj} := \lambda_{t,n}^k r_t^k / (\lambda_{t,n}^k r_t^k + \lambda_{t,n}^j r_t^j)$, we obtain

$$r_{t+1}^k [1 + \rho \sum_{n=1}^{N} (\lambda_{t+1,n}^j - \lambda_{t+1,n}^k) C_{t,n}^{kj}] = \sum_{n=1}^{N} [\rho\lambda_{t+1,n}^j + (1-\rho) R_{t+1,n}] C_{t,n}^{kj}.$$

Thus

$$\frac{r_{t+1}^k}{r_{t+1}^j} = \frac{A_{t+1}^{kj} / B_{t+1}^{kj}}{A_{t+1}^{jk} / B_{t+1}^{jk}},$$

where

$$A_{t+1}^{kj} := \sum_{n=1}^{N} [\rho\lambda_{t+1,n}^{j} + (1-\rho)R_{t+1,n}]C_{t,n}^{kj},$$

$$B_{t+1}^{kj} := 1 + \rho\sum_{n=1}^{N}(\lambda_{t+1,n}^{j} - \lambda_{t+1,n}^{k})C_{t,n}^{kj}.$$

Observe that $B_{t+1}^{jk} = B_{t+1}^{kj}$. Indeed,

$$B_{t+1}^{kj} - B_{t+1}^{jk} = \rho\sum_{n=1}^{N}[(\lambda_{t+1,n}^{j} - \lambda_{t+1,n}^{k})C_{t,n}^{kj} - (\lambda_{t+1,n}^{k} - \lambda_{t+1,n}^{j})C_{t,n}^{jk}] =$$

$$\rho\sum_{n=1}^{N}(\lambda_{t+1,n}^{j} - \lambda_{t+1,n}^{k}) = 0$$

because $C_{t,n}^{kj} + C_{t,n}^{jk} = 1$. Consequently,

$$\frac{r_{t+1}^{1}}{r_{t+1}^{2}} = \frac{A_{t+1}^{12}}{A_{t+1}^{21}} = \frac{r_t^1}{r_t^2}\frac{\sum_{n=1}^{N}[\rho\lambda_{t+1,n}^{2} + (1-\rho)R_{t+1,n}]\frac{\lambda_{t,n}^{1}}{\lambda_{t,n}^{1}r_t^1/r_t^2 + \lambda_{t,n}^{2}}}{\sum_{n=1}^{N}[\rho\lambda_{t+1,n}^{1} + (1-\rho)R_{t+1,n}]\frac{\lambda_{t,n}^{2}}{\lambda_{t,n}^{1}r_t^1/r_t^2 + \lambda_{t,n}^{2}}},$$

which yields (21.7).

Question 21.6 (The "Domination Paradox") Suppose there are two investors $k = 1, 2$ and two assets $n = 1, 2$, and one of the assets, say the first ($n = 1$), always yields relative dividends greater than the second ($n = 2$): $R_1 > R_2$, i.e. the former strictly *dominates* the latter. It might seem natural to invest in all circumstances into the strictly dominant asset. However, the results of Chap. 20, in particular, Theorem 20.3, state that the strategy Λ^* will perform better than this one and moreover, the Λ^*-user will drive the rival out of the market. Explain this seeming paradox and illustrate the situation by a numeric example.

Answer The intuitive argument in support of investing into the dominant asset appeals to a rather frequent misunderstanding of the fact that the performance of a strategy does not depend solely on the strategy itself. In fact it depends on the whole strategy profile, and in case of two investors, on the strategy of the rival. Let us demonstrate this in the following numeric example.

Assume that there is no uncertainty in the model. Asset 1 always pays relative dividend $2/3$ and asset 2 always pays $1/3$. Assume that $\rho = 1/2$. Suppose investor 1 uses the strategy Λ^* prescribing to invest in assets 1 and 2 in the proportions $2/3$ and $1/3$, while investor 2 invests all wealth in asset 1. In other words, suppose that they use the constant proportions strategies $(2/3, 1/3)$ and $(1, 0)$, respectively. Then the dynamics of the ratio $z_t = r_t^1 / r_t^2$ of their market shares will be, according to (21.7), as follows:

$$z_{t+1} = z_t \frac{(5/6)\dfrac{2/3}{(2/3)z_t + 1} + (1/6)\dfrac{1}{z_t}}{\dfrac{2/3}{2/3z_t + 1}} = z_t \frac{\dfrac{10}{2z_t + 3} + \dfrac{1}{z_t}}{\dfrac{12}{2z_t + 3}}$$

$$= \frac{10z_t + 2z_t + 3}{12} = \frac{12z_t + 3}{12} = z_t + \frac{1}{4},$$

which implies that $z_t = r_t^1 / r_t^2 \to \infty$, i.e., investor 1 drives the rival out of the market.

Mathematical Appendices

Facts from Linear Algebra

A

The material of this Appendix can be found in any comprehensive text on Linear Algebra, see e.g. Lang, S., Linear Algebra, Undergraduate Texts in Mathematics, Springer, 2004 (third edition).

Symmetric and Positive Definite Matrices A matrix $V = (v_{ij})$ is called *symmetric* if

$$v_{ij} = v_{ji} \text{ for all } i, j.$$

A matrix V is called *positive definite* if

$$\langle x, Vx \rangle > 0 \text{ for all } x \neq 0.$$

It is called *positive semidefinite* if

$$\langle x, Vx \rangle \geq 0 \text{ for all } x \neq 0.$$

The notions of a *negative definite* and *a negative semidefinite* matrix are defined analogously (with the opposite inequalities).

We denote by e_i the vector

$$e_i = (0, \ldots, 0, 1, 0, \ldots 0)$$

whose coordinates are equal to zero except for the ith coordinate which is equal to 1.

Proposition 1 *A matrix V is symmetric if and only if*

$$\langle x, Vy \rangle = \langle Vx, y \rangle.$$

I.V. Evstigneev et al., *Mathematical Financial Economics*, Springer Texts in Business and Economics, DOI 10.1007/978-3-319-16571-4

Proposition 2 *Let V be a matrix and let $W = V^{-1}$ be its inverse. If V is symmetric, then W is symmetric.*

Proposition 3 *Let V be a matrix and let $W = V^{-1}$ be its inverse. If V is positive definite, then W is positive definite.*

Proposition 4 (Cauchy–Schwartz Inequality) *If V is a positive definite symmetric matrix, then*

$$\langle x, Vx\rangle\langle y, Vy\rangle > \langle x, Vy\rangle^2$$

for all vectors x and y which are not collinear.

Remark Observe that the above inequality turns into equality if $y = \lambda x$ (or $x = \mu y$), i.e., the vectors x and y are collinear.

Orthogonal Vectors Two vectors x and y are called *orthogonal* if $\langle x, y\rangle = 0$. A linear equation with N unknowns $x_1, \ldots, x_N$,

$$a_1 x_1 + \ldots + a_N x_N = 0,$$

can be written as

$$\langle a, x\rangle = 0,$$

where

$$a = (a_1, \ldots, a_N), \; x = (x_1, \ldots, x_N).$$

A system of L linear equations with N unknowns can be represented as

$$\langle a^l, x\rangle = 0, \; l = 1, 2, \ldots, L, \tag{A.1}$$

which means that the vector x—a solution to this system—is orthogonal to each of the vectors $a^1, \ldots, a^L$.

We say that a vector b is a *linear combination* of $a^1, \ldots, a^L$ if there exist numbers $\lambda_1, \ldots, \lambda_L$ such that

$$b = \lambda_1 a^1 + \ldots + \lambda_L a^L.$$

We use in this textbook the following fact.

Proposition 5 *Let $a^1, \ldots, a^L$ and b be N-dimensional vectors. If any vector x which is orthogonal to each of the vectors $a^1, \ldots, a^L$ is orthogonal to b, then b*

is a linear combination of $a^1, \ldots, a^L$*. In other words, if any solution* x *to the system of linear equations (A.1) is a solution to the linear equation*

$$\langle b, x \rangle = 0,$$

then b *is a linear combination of* $a^1, \ldots, a^L$*.*

A set $\mathscr{L}$ of vectors in an N-dimensional space R^N is called a *linear space* if for any $x, y \in \mathscr{L}$ and any numbers λ_1, λ_2, we have $\lambda_1 x + \lambda_2 y \in \mathscr{L}$. A vector z is said to be orthogonal to $\mathscr{L}$ if $\langle z, x \rangle = 0$ for all $x \in \mathscr{L}$.

Proposition 6 *If a linear space* $\mathscr{L}$ *does not coincide with* R^N*, then there exists a vector* $z \in R^N$ *orthogonal to* $\mathscr{L}$*.*

Partial Derivatives of the Functions $\langle a, x \rangle$ and $\langle x, Vx \rangle$ We first observe that the partial derivative of the function

$$\langle a, x \rangle = a_1 x_1 + \ldots + a_N x_N$$

with respect to x_k is equal to a_k:

$$\frac{\partial}{\partial x_k} \langle a, x \rangle = a_k.$$

Consequently, the gradient (the vector of the partial derivatives) of $\langle a, x \rangle$ is given by

$$grad \, \langle a, x \rangle = a.$$

Proposition 7 *For any symmetric matrix* V*, we have*

$$grad \, \langle x, Vx \rangle = 2Vx.$$

The matrix of the second partial derivatives

$$\left(\frac{\partial^2}{\partial x_k \partial x_i} F(x) \right)$$

of a function $F(x) = F(x^1, \ldots, x^N)$ is called the *Hessian* of the function.

Proposition 8 *The Hessian of the function* $\langle x, Vx \rangle$ *is equal to* $2V$*.*

Convexity and Optimization

B

In this appendix, we give the main definitions related to the notions of convexity and concavity. We consider constrained optimization problems involving the maximization of concave functions on convex sets. We focus on problems of two types: with equality constrains and with inequality constraints. The material presented in this Appendix is covered, e.g., in D.G. Luenberger, Optimization by Vector Space Methods, Wiley (1997).

Convexity and Concavity

Definition 1 A set X in R^N is called *convex* if for any vectors x, y in X all the vectors of the form

$$z_\theta = \theta x + (1-\theta)y, \; 0 < \theta < 1,$$

belong to X.

Definition 2 A function $f(x)$ defined on X is called *concave* if

$$f(z_\theta) \geq \theta f(x) + (1-\theta) f(y) \tag{B.1}$$

for all $0 < \theta < 1$. The function $f(x)$ is called *convex* if

$$f(z_\theta) \leq \theta f(x) + (1-\theta) f(y) \tag{B.2}$$

for all $0 < \theta < 1$. If the inequalities in (B.1) or (B.2) are strict for $x \neq y$, the function is called *strictly concave* or *strictly convex*, respectively.

I.V. Evstigneev et al., *Mathematical Financial Economics*, Springer Texts in Business and Economics, DOI 10.1007/978-3-319-16571-4

Basic Facts Regarding Convex and Concave Functions

Proposition 1 *If a strictly concave function attains its maximum on a convex set, then the point of the maximum is unique. If a strictly convex function attains its minimum on a convex set, then the point of the minimum is unique.*

Proposition 2 *Let $f(x)$ be a function on R^N having continuous second partial derivatives $\frac{\partial^2}{x_i x_j} f(x)$. The function $f(x)$ is concave if and only if its Hessian, the matrix of second partial derivatives*

$$H(x) = (\frac{\partial^2}{\partial x_i \partial x_j} f(x)),$$

is negative semidefinite for each x. If $H(x)$ is negative definite for each x, then $f(x)$ is strictly concave. Analogous statements (with "positive" in place of "negative") hold for convex and strictly convex functions.

Proposition 3 *Let $f(x)$ be a concave function on R^N with continuous partial derivatives $\frac{\partial}{\partial x_i} f(x)$. Denote by $f'(x)$ the gradient*

$$f'(x) = (\frac{\partial}{\partial x_1} f(x), \ldots, \frac{\partial}{\partial x_N} f(x))$$

of $f(x)$. The function $f(x)$ attains its maximum on R^N at a point $\bar{x}$ if and only if $\bar{x}$ is its stationary point, i.e.,

$$f'(\bar{x}) = 0.$$

The analogous assertion is valid for points of a minimum of convex functions.

Optimization with Equality Constraints Let $f(x)$ be a concave function defined on R^N and let $g(x)$ be a linear function,

$$g(x) = a_1 x_1 + \ldots + a_N x_N + b,$$

where $x = (x_1, \ldots, x_N)$. Consider the following optimization problem:
($\mathscr{P}_1$) Maximize $f(x)$ on R^N subject to

$$g(x) = 0. \tag{B.3}$$

Theorem 1 *Let $\bar{x}$ be a point in R^N such that $g(\bar{x}) = 0$. Then the following assertions are equivalent:*

(i) $\bar{x}$ *is a solution to problem* $(\mathscr{P}_1)$;
(ii) *there exists a number* λ *such that* $\bar{x}$ *maximizes*

$$L(x, \lambda) = f(x) + \lambda g(x)$$

on R^N.

The function $L(x, \lambda)$ is called the *Lagrangian* and the number λ is called a *Lagrange multiplier* corresponding to constraint (B.3).

Remark 1 If $f(x)$ has continuous partial derivatives,

$$\frac{\partial}{\partial x_i} f(x),$$

then $\bar{x}$ maximizes the Lagrangian $L(x, \lambda)$ with respect to x if and only the gradient of the Lagrangian,

$$L'_x(x, \lambda) = (\frac{\partial}{\partial x_1} L(x, \lambda), \ldots, \frac{\partial}{\partial x_N} L(x, \lambda)),$$

is equal to zero at $\bar{x}$ (see Proposition 3 above).

Optimization with Inequality Constraints Let $g_1(x), \ldots, g_m(x)$ and $f(x)$ be concave functions defined on a convex set X in R^N. Consider the following optimization problem:

$(\mathscr{P}_2)$ Maximize $f(x)$ on X subject to

$$g_1(x) \geq 0, \ldots, g_m(x) \geq 0. \tag{B.4}$$

We will assume that the set X contains an element y such that

$$g_1(y) > 0, \ldots, g_m(y) > 0$$

(*Slater's condition*). Define

$$g(x) = (g_1(x), \ldots, g_m(x)).$$

Theorem 2 (Kuhn-Tucker Theorem) *Let* $\bar{x}$ *be a point in* X *satisfying (B.4). Then the following assertions are equivalent:*

(i) $\bar{x}$ *is a solution to problem* $(\mathscr{P}_2)$;
(ii) *there exists a vector* $\lambda = (\lambda_1, \ldots, \lambda_m)$ *with non-negative coordinates such that* $\bar{x}$ *maximizes*

$$L(x, \lambda) = f(x) + \langle \lambda, g(x) \rangle$$

on X and

$$\langle \lambda, g(\bar{x}) \rangle = 0. \tag{B.5}$$

The last equality is termed the *complementary slackness condition.*

Remark 2 Slater's condition is redundant (Theorem 2 holds without it) if the functions $f(x)$ and $g_j(x)$ are linear and the set X is defined by a system of linear inequalities

$$h_1(x) \geq 0, \dots, h_k(x) \geq 0,$$

where $h_1(x), \dots, h_k(x)$ are linear functions. A set X consisting of points satisfying a system of linear inequalities is called a *polyhedral set*. Under the assumptions imposed, the problem $(\mathscr{P}_2)$ is a *linear programming* problem.

In this textbook we apply Theorem 2 three times: in Chaps. 2, 5 and 12.

In Chap. 2 (Proposition 2.1) and in Chap. 5 (Proposition 5.1), to verify Slater's condition we have to construct a normalized portfolio whose expected return is strictly greater than some given number μ. To this end consider those assets i and j for which $m_i \neq m_j$ (say, $m_i < m_j$), take some number θ and consider a portfolio whose positions x_i and x_j are equal to $1 - \theta$ and θ, while all the other positions are 0. This is a normalized portfolio, and its expected return is equal to

$$(1 - \theta)m_i + \theta m_j = m_i + \theta(m_j - m_i).$$

By selecting θ large enough, we can make this expression greater than any given number μ.

In Chap. 12 (Theorem 12.1), we deal with a linear programming problem, and in this case, as it is pointed out in Remark 2, the verification of Slater's condition is not needed.

Sources

Part I

The material of *Part I* is based on, and may be used as an introduction to, the following texts:

Luenberger, D. (1998). *Investment science*. (1st ed.), 2013 (2nd ed.). Oxford: Oxford University Press.

Panjer, H. H. (Ed.) (1998). *Financial economics*. Schaumburg, IL: The Actuarial Foundation of the USA.

Werner, J., & LeRoy, S. F. (2000). *Principles of financial economics*. Cambridge: Cambridge University Press.

Part II

The main sources for *Chaps. 11–13* are as follows:

Föllmer, H., & Schied, A. (2002). *Stochastic finance: An introduction in discrete time*. Berlin: Walter deGruyter .

Pliska, S. R. (1997). *Introduction to mathematical finance: Discrete time models*. New Jersey: Wiley-Blackwell.

In *Chaps. 14 and 15*, we follow closely

Ross, S. (1999). *An introduction to mathematical finance*. Cambridge: Cambridge University Press.

Part III

The content of *Chap. 19* is well covered (with problems and exercises) by numerous textbooks. We used the following:

Magill, M., & Quinzii, M. (2002). *Theory of incomplete markets*. Cambridge: MIT Press.

This book contains proofs of all the assertions in *Chap. 19*.

Regarding other chapters in Part III, we refer to two books and several research papers, where the reader can find further references.

Chaps. 17 and 18: capital growth theory.

It might seem at first glance surprising, but the main ideas of capital growth theory were developed within mathematical theory of information, rather than economics or finance. Financial growth models and concepts are akin to those in information theory, that was created in the works of Shannon, Breiman, Kelly, Cover

I.V. Evstigneev et al., *Mathematical Financial Economics*, Springer Texts in Business and Economics, DOI 10.1007/978-3-319-16571-4

and others in 1950s–1980s. The following textbook includes chapters on growth-optimal investments:

Cover, T. M., & Thomas, J. A. (2006). *Elements of information theory* (2nd Ed.). New York: Wiley.

The volume cited below contains papers on various aspects of this subject and a detailed discussion of its history.

MacLean, L. C., Thorp, E. O., & Ziemba, W. T. (Eds.), (2011). *The Kelly capital growth investment criterion: Theory and practice. World scientific handbook in financial economics series* (Vol 3). Singapore: World Scientific.

The phenomenon of volatility-induced growth was discovered (in a specialized model) by Fernholz and Shay:

Fernholz, E. R., & Shay, B. (1982). Stochastic portfolio theory and stock market equilibrium. *Journal of Finance, 37*, 615–624.

This phenomenon was for the first time systematically analyzed in the most general setting in

Dempster, M. A. H., Evstigneev, I. V., & Schenk-Hoppé, K. R. (2007) Volatility-induced financial growth. *Quantitative Finance, 7*, 151–160.

The model considered in *Chap. 20* was proposed in

Evstigneev, I. V., Hens, T., & Schenk-Hoppé, K. R. (2006). Evolutionary stable stock markets. *Economic Theory, 27*, 449–468.

For proofs of Theorems 20.1–20.3, see:

Evstigneev, I. V., Hens, T., & Schenk-Hoppé, K. R. (2008). Globally evolutionarily stable portfolio rules. *Journal of Economic Theory, 140*, 197–228.

Amir, R., Evstigneev, I. V., Hens, T., & Xu, L., (2011). Evolutionary finance and dynamic games. *Mathematics and Financial Economics, 5*, 161–184.

A reference to fundamental indexing is as follows:

Arnott, R. D., Hsu, J. C., & West, J. M., (2008). *The fundamental index: A better way to invest*. New Jersey: Wiley.

Printed by Printforce, the Netherlands